The Machinery of Life

The Immune System Piercing a Bacterial Cell Wall

Our blood contains proteins that recognize and destroy invading cells. In this illustration, blood serum fills the upper half and a bacterial cell wall, under attack, fills the lower. Y-shaped antibody molecules recognize and bind to the foreign cell surface, and are in turn recognized by the six-armed protein at left. This begins a process that leads to the formation of an attack complex, seen piercing the cell wall at right center. These holes upset the balance of ions in the invading bacterial cell, causing it to expand and rupture. (1,000,000 ×)

The Machinery of Life

David S. Goodsell

With 93 illustrations, 16 in color

Springer-Verlag

New York   Berlin   Heidelberg   London   Paris
Tokyo   Hong Kong   Barcelona   Budapest

David S. Goodsell, Ph.D.
Molecular Biology Institute
University of California
Los Angeles, CA 90024-1570
USA

Library of Congress Cataloging-in-Publication Data
Goodsell, David S.
    The machinery of life / David S. Goodsell.
        p.   cm.
    Includes bibliographical references and index.
    ISBN 0-387-97846-1. — ISBN 3-540-97846-1
    1. Cells.   2. Molecular biology.   3. Cells-Pictorial works.
    I.   Title.
        [DNLM: 1. Cells.   2. Molecular Biology.   QH581.2 G655m]
    QH581.2.G66   1992
    574.87 — dc20
    DNLM/DLC
    for Library of Congress   92-2303

Printed on acid-free paper.

"I will put Chaos into fourteen lines" by Edna St. Vincent Millay. From *Collected Poems*, Harper & Row. Copyright © 1954, 1982 by Norma Millay Ellis. Reprinted by permission of Elizabeth Barnett, literary executor.

Acquiring Editor: Robert Garber.
Production managed by Ellen Seham; manufacturing supervised by Rhea Talbert.
Typeset by Bytheway Typesetting Services, Inc., Norwich, NY.
Color: separations of the insert by York Graphic Services, Inc., York, PA; printed by New England Book Components, Hingham, MA.
Printed and bound by Arcata Graphics/Kingsport Press, Kingsport, TN.
Printed in the United States of America.

9 8 7 6 5 4 3

ISBN 0-387-97846-1 Springer-Verlag New York Berlin Heidelberg
ISBN 3-540-97846-1 Springer-Verlag Berlin Heidelberg New York

for my parents,
David and Cheryl Goodsell

I will put Chaos into fourteen lines
And keep him there; and let him thence escape
If he be lucky; let him twist, and ape
Flood, fire, and demon—his adroit designs
Will strain to nothing in the strict confines
Of this sweet Order, where, in pious rape,
I hold his essence and amorphous shape,
Till he with Order mingles and combines.
Past are the hours, the years, of our duress,
His arrogance, our awful servitude:
I have him. He is nothing more nor less
Than something simple not yet understood;
I shall not even force him to confess;
Or answer. I will only make him good.

EDNA ST. VINCENT MILLAY

Preface

Imagine that we had some way to look directly at the molecules in our bodies, perhaps with an x-ray microscope or an Asimov-style shrinking-and-enlarging machine (unfortunately, neither is currently feasible). Think of the wonders we could witness firsthand: antibodies attacking a virus, electrical impulses shooting down nerve fibers, proteins building new strands of DNA. Many of the questions puzzling the current cadre of biochemists would be answered at a glance. But the microscopic world of molecules is separated from our everyday world by an insurmountable, million-fold difference in size.

I created the illustrations in this book to help bridge this gulf and allow us to look at the molecular structure of cells, if not directly, then in an artistic rendition. I have drawn two types of illustrations with this goal in mind: drawings that magnify a small portion of a living cell by one million times, showing the arrangement of molecules inside (as in the Frontispiece), and computer-generated pictures showing individual molecules in atomic detail (as in the color plates).

As I worked on the illustrations, several themes evolved which I have incorporated into the book. One is that of scale. Most of us do not have a good concept of the relative sizes of water molecules, proteins, ribosomes, bacteria, and people. To improve this understanding, I have drawn all of the illustrations at a few consistent magnification factors. The views showing the interiors of living cells, enclosed in squares in Chapters 4 to 7, are all drawn at one million times magnification. They may be directly compared to one another: by flipping between pages in these chapters, you can compare the sizes and shapes of DNA, lipid bilayers, nuclear pores, and the many other molecules in living cells. The computer-generated figures of individual molecules are drawn at two consistent scales: ten million times magnification to show entire macromolecules and thirty million times magnification to more effectively show small molecules.

I have also drawn the illustrations in a consistent style, again to allow ready comparison. A space-filling representation is used for all illustrations of mole-

cules, without capricious switches from ball-and-stick pictures to bond diagrams to ribbon diagrams to chemical symbols. Throughout, I have attempted
to give a consistent feel for the shape and size of the molecules.

In the drawings of cellular interiors, I have made every effort to include the
proper number of molecules, in the proper place, and having the proper size
and shape. The published data on the distribution and concentration of molecules are far from complete, however, and are continually being extended
and refined. As a result, the cellular pictures are subject to some personal
interpretation, especially in the illustrations of yeast and human cells in Chapters 5 and 6. In addition, many of the individual molecules have not yet been
analyzed at the atomic level, allowing some additional breadth for artifice.

I have written the text with the nonscientist reader in mind; I have drawn
the illustrations at a level of scientific rigor to satisfy the biochemist. For the
lay reader this book is an introduction to biochemistry—a pictorial overview
of the molecules that orchestrate the processes of life. It is far from comprehen-
sive. The reader is referred to the excellent textbooks listed at the end of the
book for more detailed information (Lubert Stryer's *Biochemistry* is an excellent place to start, and *Molecular Biology of the Cell* by Alberts et al. will
point the way to the original reports of nearly any biochemical subject). For
the biochemist, it is my hope that this book will act as a touchstone for
intuition. Please use the illustrations, as I have, to help imagine biological
molecules in their proper context: packed into living cells.

I thank the people who have been instrumental in seeing this project from
concept to conclusion. Arthur Olson provided useful comments at every stage,
as well as providing an excellent working environment in the Molecular Graph-
ics Laboratory at the Scripps Research Institute in La Jolla. The computer-
generated illustrations were produced with programs I developed under the
auspices of the Daymon Runyon–Walter Winchell Cancer Research Fund and
the National Institutes of Health. Finally, I would like to thank Bill Grimm for
his support and confidence.

Contents

Illustrations

Atomic Coordinates

The computer-generated illustrations were calculated using atomic coordinates taken from the Brookhaven Protein Data Bank. I include here the authors of each structure file and gratefully acknowledge their sense of scientific propriety in making the results of their labors available to the public.

6ADH (alcohol dehydrogenase). Eklund H.

3CLN (calmodulin). Babu YS, Bugg CE, Cook WJ.

3CRO (cro protein-DNA). Mondragon A, Harrison SC.

3DFR (dihydrofolate reductase). Filman DJ, Matthews DA, Bolin JT, Kraut J.

2ENL (enolase). Lebioda L, Stec B.

1FC1 (immunoglobulin Fc fragment). Deisenhofer J.

1GCN (glucagon). Blundell TL, Sasaki K, Dockerill S, Tickle IJ.

1GD1 (glyceraldehyde-3-phosphate dehydrogenase). Skarzynski T, Moody PCE, Wonacott AJ.

2GLS (glutamine synthetase). Eisenberg D, Almassy RJ, Yamashita MM.

1HCO (carbonmonoxy hemoglobin). Baldwin JM.

2HHB (deoxy hemoglobin). Fermi G, Perutz MF.

1HHO (oxy hemoglobin). Shaanan B.

1HKG (hexokinase). Bennett WS Jr, Steitz TA.

4HVP (HIV-1 protease). Miller M, Schneider J, Sathyanarayana BK, Toth MV, Marshall GR, Clawson L, Selk L, Kent SBH, Wlodawer A.

2IG2 (immunoglobulin G1). Marquart M, Huber R.

4INS (insulin). Dodson GG, Dodson EJ, Hodgkin DC, Isaacs NW, Vijayan M.

5PEP (pepsin). Cooper JB, Khan G, Taylor G, Tickle IJ, Blundell TL.

1PFK (phosphofructokinase). Shirakihara Y, Evans PR.

1PGI (glucose-6-phosphate isomerase). Muirhead H.

3PGK (phosphoglycerate kinase). Shaw PJ, Walker NP, Watson HC.

3PGM (phosphoglycerate mutase). Campbell JW, Hodgson GI, Warwicker J, Winn SI, Watson HC.

1PRC (photosynthetic reaction center). Deisenhofer J, Epp O, Miki K, Huber R, Michel H.

1PYK (pyruvate kinase). Muirhead H, Levine M, Stammers DK, Stuart DI.
4TNA (transfer RNA). Jack A, Ladner JE, Klug A.
3TRA (transfer RNA). Westhof E, Dumas P, Moras D.
2YPI (triose phosphate isomerase). Lolis E, Petsko GA.

The Machinery of Life

Chapter 1

Introduction

You and I are directly related to every other living thing on the earth. With our closest relatives—our parents, brothers, and sisters—we need only take a casual look to see the family resemblance. The same casual glance will show my close relationship to you, and our close kinship to all other men and women. We are all negligibly different, varying only in nuances of proportion and subtleties of shade.

Vertebrate animals—mammals, birds, reptiles, amphibians, and fish—form our extended family. They are cousins many times removed, so a short study of anatomy is necessary to see the resemblance. Our internal anatomy is virtually identical to that of other vertebrates. We all have a nervous system controlled by a brain, directing muscles to move a skeleton, and responding to input from eyes and ears. The differences between us and an elephant or a lizard are trifles of magnitude: longer legs, denser fur, or sharper teeth.

A closer look reveals our kinship with simpler animals—sponges, insects, flatworms—as well as with all plants, fungi, and protozoa. Through the microscope, we find that all are composed of cells. The cells making up our bodies have much in common with the cells of other organisms. They all have similar mechanisms for capturing food, for insulating themselves, for generating usable energy, and for reproducing.

An even closer look reveals our familial link to our distant relatives, the bacteria. The techniques of biochemistry have uncovered the fundamental similarity—the familial similarity—between these simple cells and all other life on earth: the similarity of the many molecules performing the functions of living.

Following our family tree back 3.5 to 4 billion years, we can make an informed guess about what our most distant ancestor was like. A single kind of molecule is thought to have provided the seed for the development of life. One special property set them apart from all other molecules: they could direct the formation of an identical copy of themselves using only the re-

1

sources available in their surroundings. These molecules, although arguably not alive, stumbled upon a process central to all modern life: the ability to reproduce.

Under the pressure of natural selection, these reproducing molecules evolved, gradually improving and combining with other molecules to form the first organisms. Molecular machines were discovered that perform chemical reactions, allowing organisms to make the molecules they need instead of relying on only those molecules available around them. Molecular motors were developed, providing the means for directed motion. Molecular skins were discovered, allowing organisms to sequester their resources away from the ravages of the environment.

Gradually, the organisms that were the best at reproducing dominated, eliminating their less productive competitors. The organisms that won this contest are our direct ancestors: primordial cells composed of a diverse collection of individual molecular machines, each performing one particular task in the process of life. The same specialized molecules are used today by all modern organisms—improved and streamlined, but still basically in the same form as used by the first cells. Over the long history of the earth, the secret of animating the inanimate has not been stumbled upon a second time, so we and all other living things can count this early cell as our common ancestor, the root of our family tree.

In this book we will explore our common birthright of molecular machines. Part I begins with a close look at the individual molecular machines built and used by modern cells: proteins, nucleic acids, lipids, and polysaccharides. We will then see how these molecular machines work in concert to perform the functions of life: reproduction, utilization of energy resources, and dealing with a hostile environment.

Part II puts all of this information together in illustrations of how four different organisms are arranged at the molecular level. Starting with a simple organism, the bacterium *Escherichia coli*, we will see how a very few basic types of molecules are used in many variations to perform all the tasks of living. We will then see how the same molecular machinery is repackaged, but not substantially changed, in a more complex organism: baker's yeast. Then we will look at two multicellular organisms, ourselves and a plant, to see the incredible diversity of function that is possible using only variations of the molecular machines in bacteria.

The book will conclude in Part III with a few special topics. After seeing how some of the cells in our bodies work, we will examine the need for certain vitamins, see how viruses hijack the machinery of our cells to secure their own reproduction, and explore the subtle difference between a drug and a poison.

THE WORLD OF CELLS

Cells are small, but not unimaginably so. Cells are about one thousand times smaller in length than objects in our everyday world (Fig. 1.1, 1.2). Typical cells in our bodies are about ten micrometers* in length—roughly one thousand times smaller than the last joint in your finger. A thousand-fold difference in length is not difficult to visualize: a grain of rice is about one thousand times smaller in length than the room you are sitting in. Imagine your room filled with rice grains, and this will give you an idea of the billion or so cells that make up your fingertip.

The small size of cells has a profound effect on the way they interact with

---

*1 micrometer = 1/1,000,000 meter.

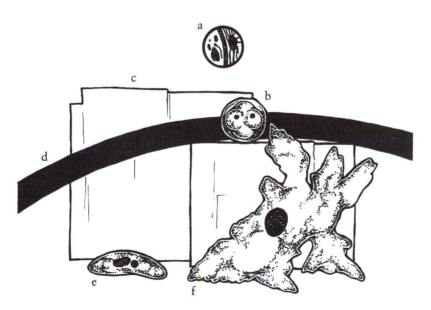

Figure 1.1   One Hundred Times Magnification

a. A collection of cells (enlarged in the next figure). b. Human egg—the largest human cell—at the four-cell stage. c. Grains of table salt. d. Human hair. e. The protozoan *Paramecium multimicronucleatum*. f. The protozoan *Amoeba proteus*.

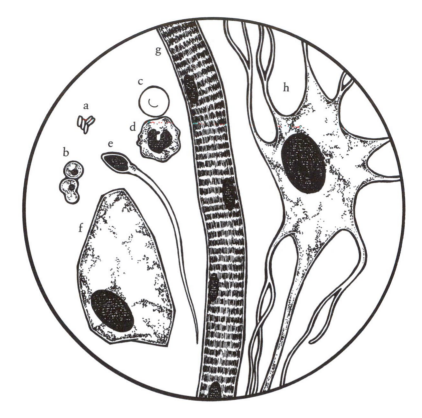

Figure 1.2   One Thousand Times Magnification

a. Five *Escherichia coli* bacteria cells. b. Two cells of baker's yeast, *Saccharomyces cerevisiae*, one in the process of budding. c. Human red blood cell. d. Human lymphocyte. e. Human sperm cell. f. Human epidermal cell. g. Human striated muscle cell. h. Human nerve cell.

their surroundings. For objects their size, gravity is not the overwhelming force that it is in our lives. The constant press of surrounding water is far more important. For objects the size of cells, water is not the flowing liquid it is to us, but is very viscous. The surface tension of water is a familiar example: small insects can sail along the surface of a pond, but if we try, the force of gravity on our enormous bodies overwhelms the gentler forces of water and we plunge to the bottom. Cells live in a world of thick, viscous water, almost oblivious to gravity. When moving from one place to another, most of their energy is spent trying to push through the gooey liquid, not in lifting their weight up from the ground.

THE WORLD OF MOLECULES

Another one thousand times reduction in size takes us from the world of cells into the world of molecules (Fig. 1.3, 1.4). An average protein used by the cell, made of about five thousand atoms, is about one-thousandth the length of a cell, or about one-millionth the length of your fingertip. If we could magnify objects by one million times, molecules would be easily visible. Atoms would be just visible, looking about the size of a grain of salt. Again, think of the rice grains packed into your room, and this will give an idea of the billion or so protein molecules that could be packed into the space of a single cell.

Molecules are so small that gravity is completely negligible: the "lives" of

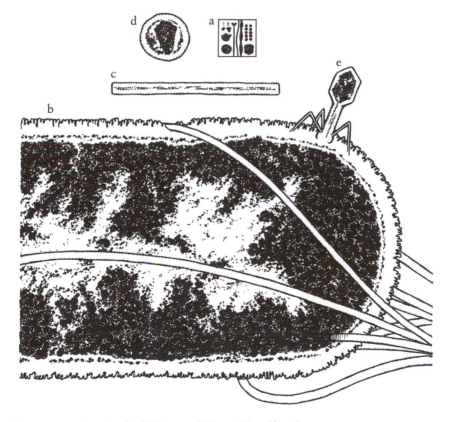

Figure 1.3   One Hundred Thousand Times Magnification

a. A collection of molecules (enlarged in the next figure). b. A bacterial cell (*a* in the previous figure). c. Tobacco mosaic virus. d. Human immunodeficiency virus. e. A bacterial virus.

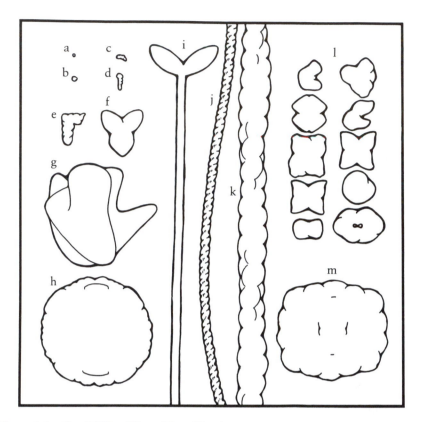

Figure 1.4   One Million Times Magnification

a. Carbon atom. b. Glucose. c. Adenosine triphosphate (ATP). d. Chlorophyll. e. Transfer RNA. f. Antibody. g. Ribosome. h. Poliovirus. i. Myosin. j. Deoxyribonucleic acid (DNA). k. Actin. l. The ten enzymes of glycolysis. m. Pyruvate dehydrogenase complex.

the large molecules in cells are completely dominated by surrounding water molecules. At room temperature, a medium-sized protein molecule travels at a rate of about five meters every second. If placed alone in space, this protein would travel a distance equal to its own length (ten nanometers) in about two nanoseconds.* In the cell, however, this protein is completely surrounded by water molecules. It is battered from all sides. It bounces back and forth, always at great speed, but never gets much of anywhere. When surrounded by water,

---

*1 nanometer = 1/1,000,000,000 meter. 1 nanosecond = 1/1,000,000,000 second.

this average protein now requires almost a thousand times as long to move the same ten-nanometer distance than if it were free to travel unimpeded.

Imagine a similar situation in our world. You enter an airport terminal and want to reach a ticket window across the room. The distance is several meters—a distance comparable to your own size. If the room is empty, the distance is crossed in a matter of seconds. But imagine instead that the room is jammed full of other people trying to get to their respective windows. With all of the pushing and shoving, it now takes you fifteen minutes to cross the room! In this time, you may be pushed all over the room, perhaps even back to your starting point a few times. This is similar (although molecules don't have a goal in mind) to the contorted path that molecules take in the cell.

MOLECULAR ILLUSTRATIONS

A measured amount of artifice is necessary in any illustrated book about molecules, as there is no way to look at them directly. What does a molecule look like? Molecules are collections of atoms, each composed of an incredibly tiny nucleus surrounded by a cloud of electrons. They are far too small to examine directly with our eyes, even if aided by the most powerful light microscopes. However, two experimental techniques come close to allowing us a direct look at atoms and molecules. Electron microscopes use a beam of electrons, captured photographically, instead of light. The result is just what we would expect. Pictures from the electron microscope show fuzzy clouds of electrons surrounding invisible nuclei. Scanning tunneling microscopy takes a more sculptural approach. A fine probe is scanned along the surface of a molecule, touching it at many points and building up a map of one side. Whereas electron microscopes indirectly show us what molecules would *look* like, scanning tunneling microscopes indirectly show us what molecules would *feel* like.

These limited pictures are not overly useful, however, in the study of the large molecular machines used by cells. In electron micrographs, the many atoms overlap and form a confusing muddle. In scanning tunneling microscope models, only a gross, sometimes distorted, picture of the surface is obtained. Instead, we use artificial representations of molecules, drawn using experimental information on the three-dimensional placement of each of their atoms in space.

In this book, individual molecules are illustrated in two ways (Fig. 1.5). The first is a simple outline drawing that gives a rough idea of the gross shape and size of the molecule. All of the outline drawings in this book are drawn at a

a.   1,000,000×

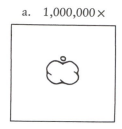

b.   10,000,000×

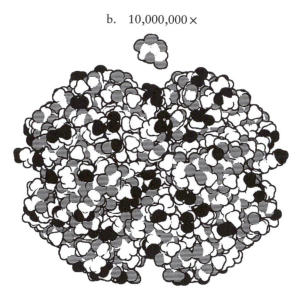

Figure 1.5   Molecular Illustrations

Hemoglobin (the protein that carries oxygen in our blood) and glucose are shown at three magnifications. The small glucose molecule is drawn over the larger hemoglobin in each. a. At one million times magnification, individual atoms are about the size of a grain of salt. Outline illustrations are used at this magnification and individual atoms are not drawn. b. At ten million times magnification, atoms are slightly smaller than a pea. All atoms are drawn in a space-filling illustration. c. At thirty million times magnification, an entire hemoglobin molecule is too large to fit on the page, but the arrangement of atoms in the glucose molecule is easily seen. Again, all atoms are drawn in a space-filling illustration.

consistent scale of one million times magnification, so the figures may be directly compared to see the relative sizes and shapes of different molecules.

A second type is a space-filling illustration that shows every atom in the molecule. For each atom, a sphere is drawn centered on the nucleus and just big enough to enclose the electrons. Space-filling pictures are useful because

c.   30,000,000×

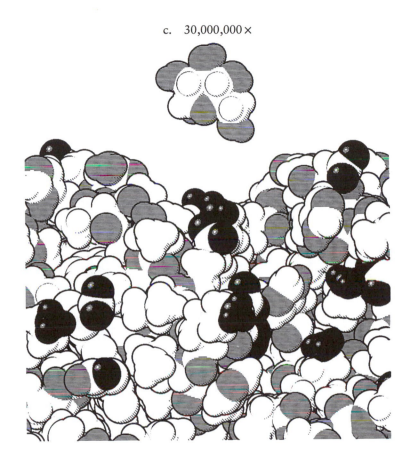

they show the location of each atom and the detailed volume of space filled by the molecule—the volume of space into which other atoms cannot penetrate. Notice that the outline drawing is just a simplification of the space-filling picture, rounding off all of the peaks and valleys. In this book, space-filling illustrations are drawn at two magnifications: ten million times for illustrations of the large molecular machines used by cells, and thirty million times for close-ups and for smaller molecules composed of fewer than one hundred atoms.

Remember, however, when looking at the drawings and computer graphic illustrations in this book, that they are simply models of fuzzy clouds of electrons surrounding collections of tiny nuclei.

I: Molecules and Life

Modern cells build the most
complex molecules on the earth—
molecular machines, each with a
specific structure, endowed with
distinct properties, tailored to a
specific functional need.

Chapter 2

Molecular Machines

Molecules orchestrate the processes of life: complex, diverse molecules with focused functional capabilities. Like the machines of our modern world, these molecules are built to perform specific functions efficiently, accurately, and consistently. Modern cells build hundreds of thousands of different molecular machines, each performing one of the hundreds of thousands of individual tasks in the process of living. These molecular machines are built according to four basic molecular plans. Whereas our macroscopic machines are built of metal, wood, plastic, and ceramic, the microscopic machines in cells are built of protein, nucleic acid, lipid, and polysaccharide. Each plan has a unique chemical personality ideally suited to a different role in the cell.

The concept of *hydrophobicity* allows us to get an intuitive feel for these chemical personalities. Because a typical cell is about 70% water by weight, the interactions of molecules with water will often dominate their behavior in the cell. Molecules tend to interact in one of two ways with water. Molecules that carry an electrical charge—usually those rich in oxygen and nitrogen atoms—interact favorably with water and are termed *hydrophilic* ("water-loving"). Hydrophilic molecules easily dissolve in water, surrounding themselves with a comfortable shell of water molecules. Table sugar and acetic acid (the sour acid in vinegar) are familiar hydrophilic molecules. On the other hand, molecules that are uncharged—usually those rich in carbon atoms—do not react favorably with water and are termed *hydrophobic* ("water-fearing"). When placed in water, hydrophobic molecules tend to cluster, pushed into globules with only a minimal outer surface in contact with water. This is what happens to oil in water: droplets form to minimize the contact of hydrophobic oil with water.

Large molecules are generally more complex than this. They often have both hydrophobic and hydrophilic regions. Different patterns of hydrophobicity cause large molecules to act in novel ways when dissolved in water. Living cells take advantage of these distinctive reactions to water in their suite of molecules.

NUCLEIC ACIDS

Nucleic acids are used to store and transmit hereditary information in cells. The unique interaction of nucleic acids with other nucleic acids makes them ideally suited to this purpose. Nucleic acids are composed of long chains of nucleotides (Fig. 2.1a). Each nucleotide is composed of a sugar-phosphate group and a disk-shaped base group. The sugar-phosphate groups link together to form a highly hydrophilic backbone (phosphates carry a negative electrical charge) and the predominantly hydrophobic bases hang off the side of this chain. When placed in water, neighboring bases stack like a roll of pennies to isolate themselves as much as possible from the surrounding water. One side of the stack is protected by the hydrophilic backbone, but the opposite edge of each base is left exposed. Herein lies the key to the utility of nucleic acids: these exposed edges have been selected over the history of life to specifically interact with other nucleic acid strands.

In DNA (deoxyribonucleic acid, one form of nucleic acid), the bases come in four varieties: adenine (A), thymine (T), cytosine (C), and guanine (G). The edges of these bases are chemically complementary: the edge of A forms two specific hydrogen bonds with the edge of T—and only with T—and the edge of C forms three specific hydrogen bonds with only G (Fig. 2.1b). (Hydrogen bonds are weaker than the bonds that hold the atoms in molecules together, but a handful can hold two molecules in tight association. The surface tension in the water, discussed in the Introduction, is due to hydrogen bonds between individual water molecules.) The matching patterns of hydrogen bonds allow a second strand of DNA—if it has the proper sequence of bases—to nestle up to the first and form a stable, two-stranded complex, with the bases happily protected from water inside.

The sequential order of bases in a strand of nucleic acid is an excellent way of storing information. In this age of computers, we have a direct analogy: the order of bases in a strand of nucleic acid is just like the string of numbers on a computer tape or disk. The numbers on computer media may store a computerized image, the directions for running a video game, or the text of this book; the bases in DNA store the information needed to make new proteins.

Nucleic acid is an especially useful molecule for storing information, because there is a direct molecular way of copying the information to a second strand. The complementary natures of the bases allow a new strand to be built using an existing strand as a template. Individual nucleotides are lined up alongside the existing strand (A specifically next to T, G specifically next to C, etc.) and connected together. The ordering of bases in the newly formed strand will be

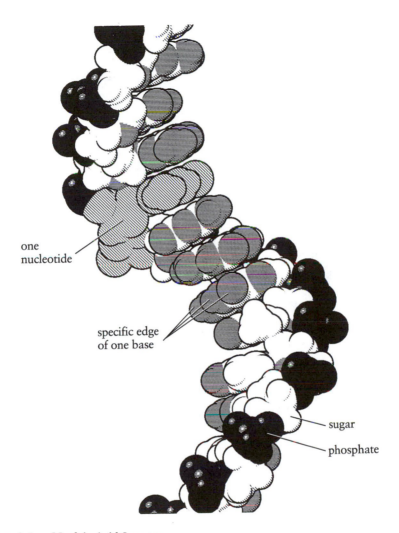

one
nucleotide

specific edge
of one base

sugar

phosphate

Figure 2.1a   Nucleic Acid Structure

Nucleic acid strands are remarkably uniform chains of similar nucleotides. The atoms in this short stretch of DNA are colored according to their hydrophobicity: hydrophobic atoms are white, slightly hydrophilic atoms are hatched, and very hydrophilic atoms are black. In this and the following figures, darker atoms prefer water. The atoms in one nucleotide building block — including one sugar, one phosphate, and one base — are highlighted with diagonal lines. (30,000,000 ×)

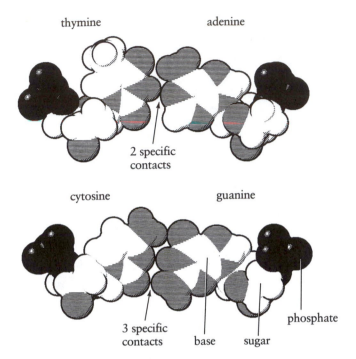

thymine   adenine

2 specific
contacts

cytosine   guanine

3 specific
contacts   base   sugar   phosphate

Figure 2.1b   Base Pairing in Nucleic Acids

Viewing single pairs of bases from a DNA double helix shows the two specific
hydrogen bonds formed when adenine pairs with thymine and the three of cytosine
with guanine. (30,000,000 ×)

completely specific: T will be found where A was in the original strand, etc.
The new "daughter" strand will contain the same information as its "parent,"
but in complementary form, as a plaster mold is complementary to the original.
A second round of duplication, using the daughter strand as a template, will
produce a sequence of bases exactly identical to the original.

The duplication of nucleic acid, all by itself, has been performed in the test
tube. Specially activated nucleotides, similar to those believed to be present in
the sea during the early history of life, dutifully line up along a single nucleic
acid strand and link together, in the expected complementary sequence, one
by one. The first "living" molecules are thought to have been just this type
of self-duplicating nucleic acid. The ability to copy information allowed the
primordial nucleic acids, as well as all modern life, to duplicate their libraries
of information and pass them on to offspring.

Nucleic acid comes in two varieties in modern cells: DNA and RNA (ribonucleic acid) (Fig. 2.2). RNA differs from DNA by having one additional oxygen atom on each sugar and one missing carbon atom on each thymine base. DNA and RNA perform different functions in modern cells. The greater stability of DNA (due to the missing oxygen on each sugar) makes it ideal as the central storehouse of information. The genetic information of each organism is stored in paired strands of DNA—the celebrated DNA double helix. The less stable RNA is used in more transient roles: as a messenger, a translator, and a synthetic machine.

Nucleic acids, however, are too limited in structure to perform the thousands of tasks in the daily life of the cell. The chemistry of the four bases is too similar to allow the construction of the varied machines needed to perform

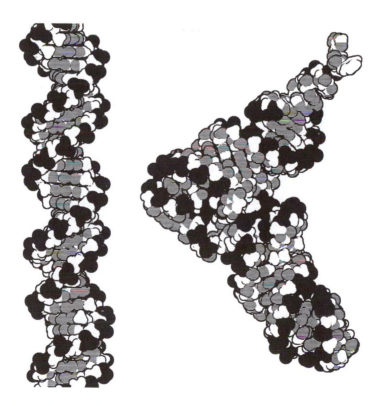

Figure 2.2   DNA and RNA

The DNA double helix on the left is composed of two strands of nucleic acid, and the transfer RNA molecule on the right is composed of a single strand folded back onto itself. (10,000,000 ×)

the thousands of necessary chemical reactions. Proteins are used in this capacity.

PROTEINS

Like nucleic acids, proteins are long, unbranched molecular chains (Fig. 2.3). Instead of using four chemically similar building blocks, however, proteins are built of twenty chemically diverse building blocks: the amino acids. Some amino acids carry an electrical charge and are hydrophilic. Others are rich in carbon and strongly hydrophobic. Some are big and bulky, others small. Some are rigid, others flexible. Some are chemically reactive, others neutral. The cell has a diverse alphabet of amino acids from which to build an even more diverse set of proteins.

The order of the hundreds or thousands of amino acids in a protein chain determines the particular shape it adopts and the function it performs (Fig. 2.4a). When a protein chain is placed in water, it twists and folds, finding the optimum shape to shelter its hydrophobic groups inside and display its hydrophilic groups outside. The final shape is completely predetermined by the order of amino acids in the chain: proteins are self-assembling machines. Proteins tend to be roughly spherical, but special sequences can fold to form long rods, nutcrackers, tubes, or a variety of other useful shapes.

The three-dimensional shape adopted by a protein in water is crucial to its function. For example, an enzyme is a protein that folds to form a cleft on one side, lined with reactive amino acids (Fig. 2.4b). The cleft is exactly the proper shape to bind to a given small molecule, such as a sugar, and position it so that the reactive amino acids can make a chemical modification, such as removing an oxygen atom. By precisely positioning a target molecule next to the proper reactive groups, enzymes specifically perform and regulate chemical reactions.

Structural proteins are designed for a different function. They fold to form a patch on one side that specifically binds to a patch on the opposite side of an adjacent molecule. This is not of much use for a single protein, but when many are placed together, they stack end to end to form the strong girders that support and move cells.

Proteins also act as carriers, folding to form pockets that hold an oxygen molecule or an atom of calcium and gently release it where needed. Other proteins form motors, turning huge molecular oars that propel bacterial cells. Many hormones are protein, messages built in one place and read like Braille in another.

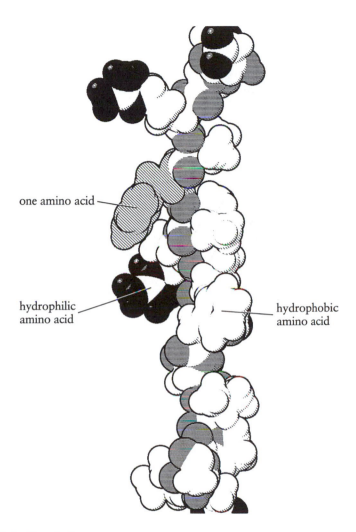

one amino acid

hydrophilic
amino acid

hydrophobic
amino acid

Figure 2.3   Protein Structure

Protein strands are heterogeneous: large, small, uncharged, and charged amino acids are seen in this short length taken from a bacterial protein. The atoms of a single amino acid are shaded with diagonal lines. (30,000,000 ×)

Proteins, with their wide potential for diversity, perform most of the everyday tasks of the cell. A typical bacterial cell builds over a thousand different types of protein, each with a different function. Our cells build about sixty thousand different kinds. Only those specific functions ideally suited to the specialized natures of nucleic acids, lipids, and polysaccharides are not per-

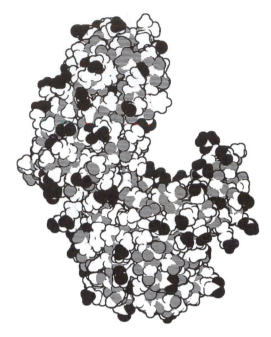

Figure 2.4a   Phosphoglycerate Kinase

Phosphoglycerate kinase is an enzyme composed of one protein chain of 415 amino acids. The chain folds into this distinctive shape, with two large lobes connected by a flexible hinge. The active site, where the chemical reaction occurs, is between the two halves. (10,000,000 ×)

formed by proteins. They are the jacks-of-all-trades, pressed to duty in innumerable shapes and forms.

LIPIDS

The unusual interaction of lipids with water makes them useful to the cell. Lipids, commonly known as fats and oils, are composed of a small hydrophilic "head" carrying two or three long hydrophobic "tails" (Fig. 2.5). When placed in water, lipid molecules spontaneously aggregate to shelter their tails. When you put grease—a typical lipid—in water, it forms small droplets with all of the tails oriented inward and all of the charged heads facing out toward the surrounding water. Lipids aggregate in cells to form a lipid bilayer, a continuous sheet composed of two layers (Fig. 2.6). The tails pack side by side at the center and the heads are displayed on the two faces, comfortable in the

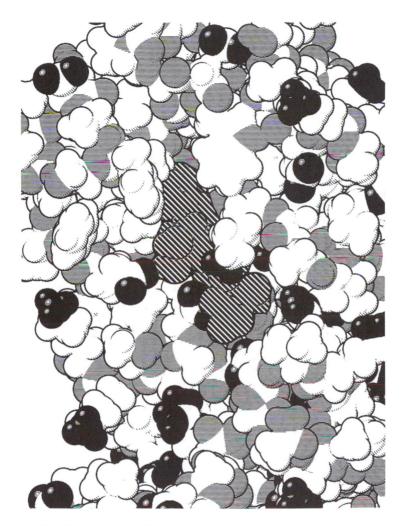

Figure 2.4b   Enzyme Active Site

The active site of phosphoglycerate kinase is shown in this close-up, with a bound molecule of ATP, darkened with diagonal lines. (This view is looking from the right in the previous figure, centered on the upper lobe.) Notice how amino acids from the enzyme wrap around and hold the ATP molecule in a specific place. (30,000,000 ×)

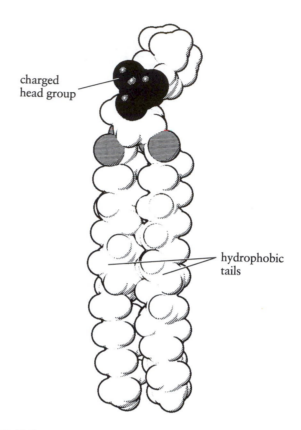

Figure 2.5   Lipid Structure

Lipids are small compared to proteins and nucleic acids, and far more hydrophobic. (30,000,000 ×)

surrounding water. Lipid bilayers are used by all cells as their primary boundary between the inside of the cell and the environment, sealing in their large molecules, such as proteins and nucleic acids.

Lipid bilayers are dynamic and fluid, because they are composed of so many separate molecules. Each individual molecule rotates about its long axis and its hydrophobic tails constantly wag and flail. Lipid molecules also slide past one another, always staying in the sheet, but randomly migrating in two dimensions.

Pure lipid bilayers are rarely found in modern cells. Proteins are inserted into most membranes, sticking into one side or completely spanning the whole sheet (Most of the figures in Part II show lipid bilayers studded with proteins.

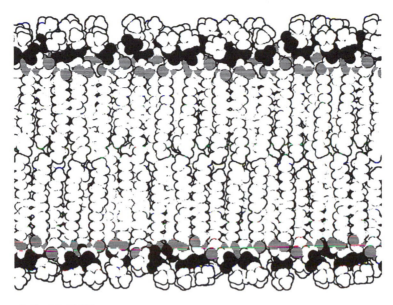

Figure 2.6   Lipid Bilayer

Many individual lipid molecules aggregate side by side to form a dynamic lipid bilayer, shown here in cross section. (10,000,000 ×)

Figure 5.4 is a particularly concentrated example.) The cell communicates through these proteins, otherwise, lipid bilayers are watertight barriers. Like the individual lipid molecules, these embedded proteins are free to move around in the fluid lipid bilayer, like icebergs in a polar sea.

POLYSACCHARIDES

Sugars, if not directly broken down for energy or used as components of nucleic acids, are linked up into long, branched chains called polysaccharides (Fig. 2.7). Unlike the order of the amino acids in proteins and the nucleotides in nucleic acids, the order of sugars in polysaccharide chains is not precisely determined. The order is not directly encoded in DNA, and polysaccharides cannot pass their sequence information on to other polysaccharides. Instead, a few enzymes repeatedly add sugars to the end of a growing chain. A different enzyme adds each different type of sugar. Except in a few specialized short chains, no two of the final polysaccharides are exactly the same. This variability is not a problem, however, for the limited uses made of polysaccharides.

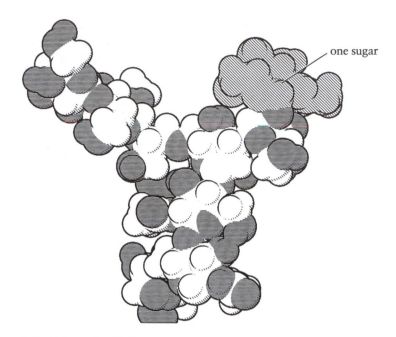

one sugar

Figure 2.7   Polysaccharide Structure

Polysaccharides such as this are attached to many proteins on the surface of animal cells. Notice that there are no strongly hydrophilic groups, making for a relatively neutral molecule. One sugar building block is highlighted with diagonal lines. (30,000,000 ×)

Polysaccharides are used as the central storehouse of energy. In lucrative times, enzymes attach any surplus sugar molecules to large granules of polysaccharide. In tougher times, other enzymes break off sugar molecules as needed. Polysaccharides are rather neutral and unreactive, making them better for storage of energy than high concentrations of free sugar. In plants, the simple sugar glucose is stored as starch, the same starch we use to thicken sauces and stiffen collars. In our cells, glucose is linked in a slightly different way to form glycogen, an oily substance.

Polysaccharides are also used in structural roles. The building you are in is probably made largely of polysaccharide: the cellulose fibers in wood are long chains of sugar. Unfortunately, they are linked together in a different way than the sugars in starch, so we cannot digest them and therefore cannot use wood or paper for food. The shells of insects are also made of a polysaccharide called chitin. Animal cells are coated with a layer of short polysaccharides, linked at one end to proteins in their cell membranes. They form a gluey coat around the cell that acts as a protective barrier. To get an idea of what this polysaccharide barrier is like, mucus owes its distinctive properties to polysaccharide chains attached to its component proteins.

Chapter 3

The Processes of Living

Imagine a leisurely walk through a wooded park. Oak and maple trees cast dancing shadows under the noonday sun. A brook gurgles off to your side. Butterflies and bees dart around colorful flowers. The wind rustles some fallen leaves on the path.

Which of the things you see are living? This seemingly trivial question is not easily answered with a formal definition. Exceptions are found to each new rule. You might say that living things can move. But the stream and the sun move, whereas the trees and the flowers are fixed in place. You might say that living things grow. But the stream also grows and dies throughout the year, and crystals can be remarkably lifelike in their growth. Looking for a more rigorous definition, you might say that living things are composed of certain amounts of carbon, oxygen, hydrogen, and nitrogen, along with a few trace elements. But the devil's advocate points to the dead leaves at your feet: they have a chemical composition very similar to living leaves but are destined only for decomposition.

This problem has about as many answers as there are biologists and philosophers interested in answering it. In this book, a descriptive definition consistent with current ideas of biochemistry will suffice: living organisms are collections of molecules that, together, perform three unusual functions. First, organisms make all of their component parts from resources available in the environment, replacing worn-out parts and reproducing. Second, organisms insulate themselves from the environment, weathering hard times or moving to more fertile grounds. Third, organisms utilize environmental sources of energy to fuel the other two processes.

Solutions to these three problems were worked out in the very early evolution of life. Molecular machines were developed and refined to perform the many individual tasks in each—machines composed of protein, nucleic acid, lipid, and sugar. In this chapter, we will see how many different molecular machines work in concert to perform these three functions.

HARNESSING AVAILABLE ENERGY

Life in our modern world is dominated by the ready availability of oxygen. Oxygen in the air is highly reactive and readily combines with carbon-rich molecules to release energy. The burning of natural gas is a familiar example: methane combines with oxygen, releasing enough energy to cook food and heat homes. Many bacterial cells and all higher cells use this supply of oxygen to generate the energy needed for all aspects of their lives. Cells consume sugars and fats, which are rich in carbon and poor in oxygen. They break them apart, combine the pieces with oxygen, and harness the released energy.

Cells cannot simply burn food like a stove: the heat would be released all at once and could not be captured and utilized. Cells take a less direct approach. Food is broken down in many small steps, adding individual oxygen atoms one by one. Because only a small change is made at each step, and because each is performed by a specific enzyme, the speed and extent of each chemical reaction is carefully controlled. In addition, the course of the breakdown has been designed to include several highly favorable steps that make ATP (adenosine triphosphate), the molecule used to transport chemical energy in the cell.

The utility of ATP lies in two unstable bonds. ATP is identical to the adenine (A) nucleotide used in RNA, except that it has three phosphate groups attached to the sugar, instead of one (Fig. 3.1). Each phosphate carries a negative electrical charge, so a string of three in close proximity is highly unstable. Energetically, it is difficult to link the three phosphates together, and, conversely, it is easy and favorable to break one or both of the phosphate–phosphate bonds.

The two unstable bonds of ATP mediate the transfer of chemical energy

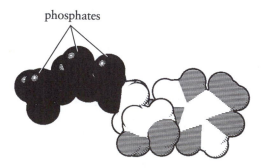

phosphates

Figure 3.1   ATP

The three charged phosphate groups of ATP (adenosine triphosphate) are held in unstable proximity. (30,000,000 ×)

from one process to another in the cell. Very favorable processes, such as the breakdown of sugar, are designed to create unstable ATP bonds. A few key enzymes in the course of the pathway perform two reactions simultaneously: a very favorable reaction, such as the addition of oxygen to carbon, is linked to an unfavorable reaction, the formation of a phosphate bond in ATP. Together, the overall reaction performed by the enzyme is slightly favorable. The coupled reaction occurs in the cell, albeit not as speedily as the reaction would without the formation of ATP. Thus, the cell makes unstable ATP bonds as it breaks down sugar.

ATP is then used to drive reactions normally impossible in the cell. Enzymes performing unfavorable reactions, like the synthesis of a protein, link their reactions with the favorable breakage of an ATP bond. Again, both reactions are performed in concert—the enzyme makes a new bond while breaking an ATP bond—yielding a reaction which is slightly favorable overall.

Looking at the cell as a whole, these processes form an ATP cycle. Phosphate bonds in ATP are created in the breakdown of food. The energy invested in these bonds is then used to fuel less favorable processes, such as synthesis of large molecules and motion. In an active cell, each ATP molecule traverses this cycle every few seconds, created and used again and again. ATP is not, however, a good molecule for storing large amounts of energy. It is far too unstable: large concentrations would break down in a short period of time. Energy is stored in neutral molecules of polysaccharide, as discussed in Chapter 2. ATP is used, instead, as the cell's currency of chemical energy.

The food consumed by most organisms is a heterogeneous mixture of proteins, fats, polysaccharides, and simple sugars. It is broken down and converted into usable energy, in the form of ATP, in three major stages.

The first stage is *digestion*, where large molecules are broken into their component pieces. Digestion funnels many different complex foods into a workable number of simple molecules. Thousands of different proteins are broken into twenty types of amino acids. Complex polysaccharides are broken into a handful of different simple sugars. Fats are broken into a few fatty acids and glycerol.

Digestion occurs largely outside of cells. Bacteria secrete enzymes into the surrounding medium, food is broken down there, and the pieces of food are absorbed through small pores in the surface of the bacteria. Special molds are cultivated because the digestive enzymes they secrete change the flavor and consistency of milk solids in pleasing ways, forming cheeses. Digestion also occurs outside our cells, but inside our bodies. The cells that line our stomach and intestines secrete enzymes and acid, and later absorb the digested food.

Digestive enzymes are generally small and very stable—destructive little automata that work under very harsh conditions. The acidity of our stomach will unfold and inactivate most proteins. Digestive enzymes are made resistant to unfolding by a number of internal crosslinks, which glue the chain into the proper conformation (see Color Plate 3). Digestive enzymes present a paradox by their very nature: proteins are made inside cells, but active digestive enzymes cannot be, because they would immediately cut up anything they touched. Cells circumvent this problem by making a protein chain which is a little too long. The longer chain folds into an inactive enzyme, with the extra bit blocking the reactive site. The cell shuttles this harmless enzyme outside, where other digestive enzymes clip off the extra amino acids and activate the enzyme.

The fragments resulting from digestion feed the second stage of energy production. The goal of the second stage is to convert the hundred or so different fragments into a single type of molecule, an acetyl group, which is shuttled to the last stage by a carrier molecule, CoA (coenzyme A). Each type of food fragment—amino acids, simple sugars, and pieces of lipids—is converted to acetyl-CoA by a different series of enzymes. A diverse set of enzymes make changes in small steps, one breaking a fatty acid in half, another adding an oxygen atom to a sugar, another discarding a nitrogen atom from an amino acid.

A central pathway in the second stage is *glycolysis* ("sugar breaking"), where glucose, a simple sugar, is converted into two molecules of acetyl-CoA in ten steps (Fig. 3.2). Each step is performed by a specific enzyme, and the seventh step is so favorable that it is used to make ATP. The logic of molecular processes is exemplified in glycolysis (making it the bane of biochemistry students): the seemingly tortuous route from start to finish is entirely necessary and ultimately quite elegant. The course of evolution has selected a pathway of small steps—allowing careful regulation of each reaction—and has designed two favorable reactions that form ATP.

In the first three steps (Fig. 3.2a-c), two phosphate "handles" are added

---→

Figure 3.2   Glycolysis

The following pages 31–40 detail the ten steps of glycolysis, the pathway of glucose degradation. The simple chemical reaction performed at each step is shown in the box (30,000,000 ×) and the specific enzyme catalyzing the reaction is shown above (10,000,000 ×). The sites of reaction on each enzyme are darkened. (Note that all atoms in the enzymes are colored white, hydrophobic and hydrophilic alike, for clarity.) The position of every atom in each enzyme has only been made publicly available by researchers in six of these enzymes. The approximate positions of each amino acid are available for three more (these pictures have a smoother outline) and no data have been made available for one (shown as an approximate outline).

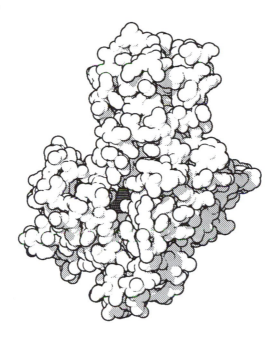

HEXOKINASE

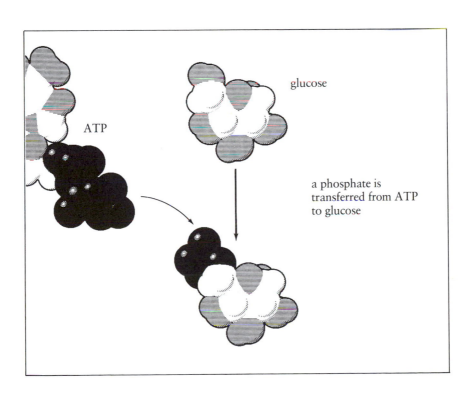

glucose

ATP

a phosphate is
transferred from ATP
to glucose

Figure 3.2a

GLUCOSE PHOSPHATE ISOMERASE

a few atoms in the sugar are rearranged, changing glucose to fructose

Figure 3.2b

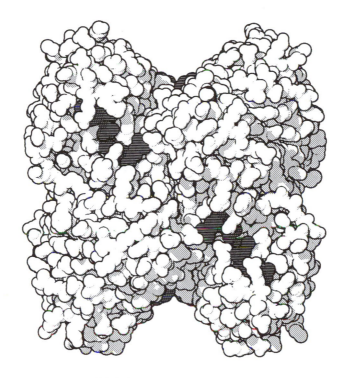

PHOSPHOFRUCTOKINASE

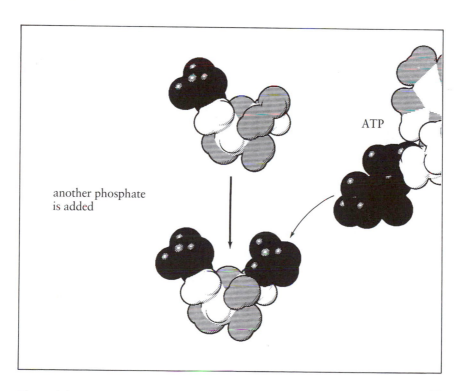

another phosphate
is added

ATP

Figure 3.2c

33

ALDOLASE

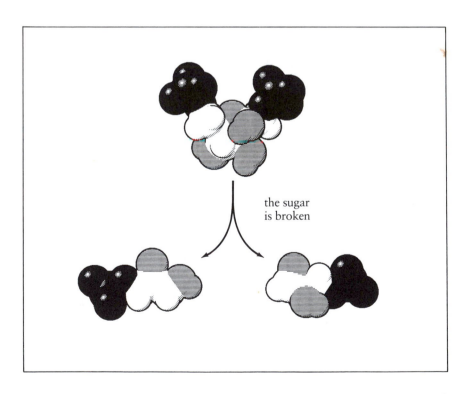

the sugar
is broken

Figure 3.2d

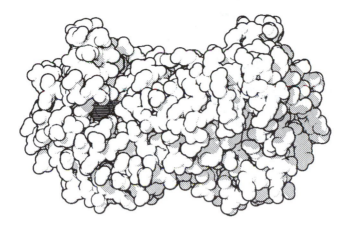

TRIOSE PHOSPHATE ISOMERASE

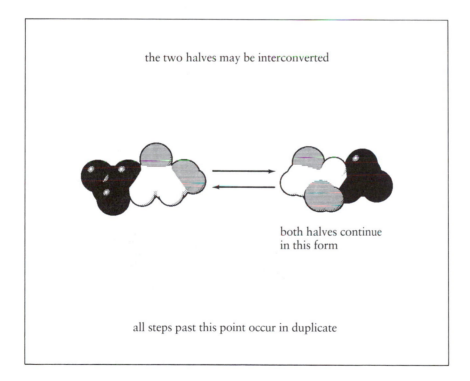

Figure 3.2e

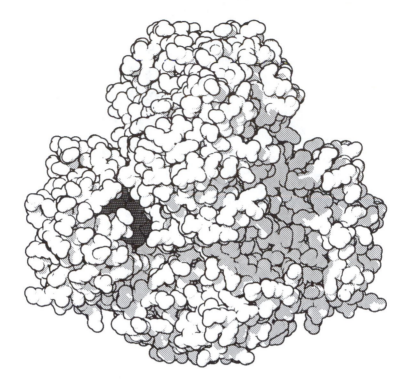

GLYCERALDEHYDE PHOSPHATE DEHYDROGENASE

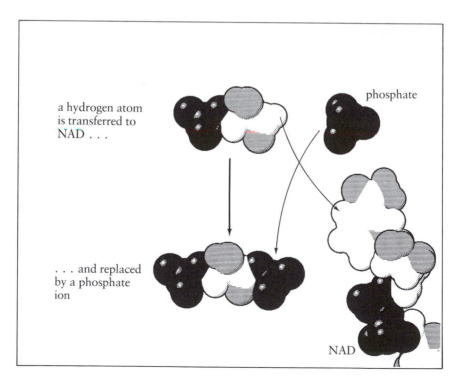

a hydrogen atom is transferred to NAD . . .

phosphate

. . . and replaced by a phosphate ion

NAD

Figure 3.2f

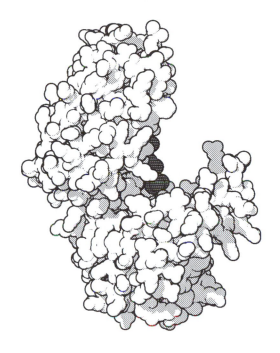

PHOSPHOGLYCERATE KINASE

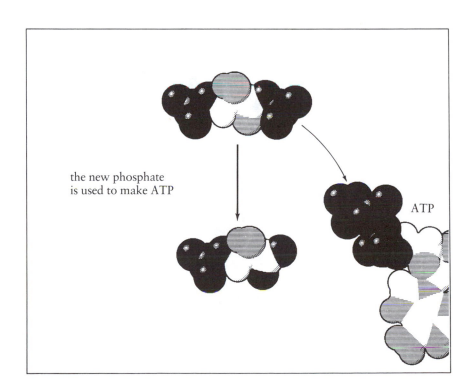

the new phosphate
is used to make ATP

ATP

Figure 3.2g

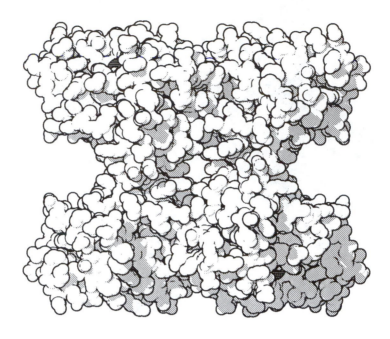

PHOSPHOGLYCEROMUTASE

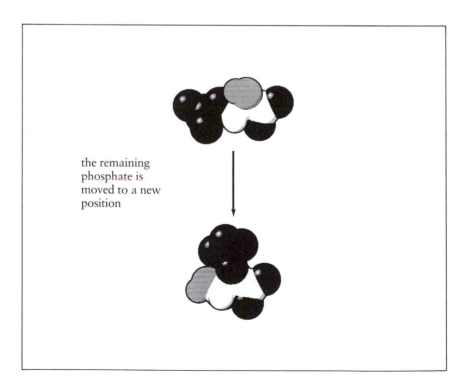

the remaining
phosphate is
moved to a new
position

Figure 3.2h

ENOLASE

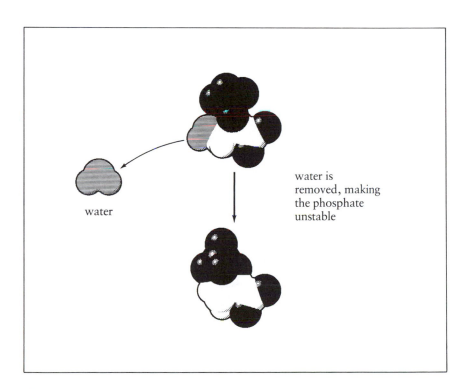

water

water is
removed, making
the phosphate
unstable

Figure 3.2i

39

PYRUVATE KINASE

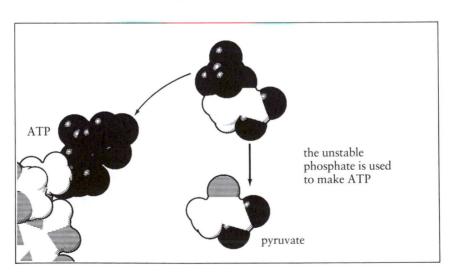

ATP

the unstable
phosphate is used
to make ATP

pyruvate

Figure 3.2j

to the ends of glucose. Phosphate bears a negative charge, making a readily identifiable marker for subsequent enzymes: the glucose molecule is now tagged for breakdown. Two ATP molecules are used to add the phosphates, priming the engine. Therefore, somewhere in the next seven steps these two ATP molecules must be restored before any net ATP is gained.

The next two steps break the molecule into two identical pieces. First a bond at the center is broken, creating two different molecules (Fig. 3.2d). Then two hydrogen atoms are rearranged on one half, making it identical to the other (Fig. 3.2e). The result is two identical halves; the remaining steps occur in duplicate, once for each half.

The sixth step (Fig. 3.2f) is the key to glycolysis. A hydrogen atom is removed and replaced by a phosphate. A free phosphate is used in the reaction, so a new phosphate–carbon bond is obtained without having to use ATP. The enzyme couples the favorable removal of hydrogen to the unfavorable, but very useful, bonding of phosphate to the molecule. (The hydrogen is transferred to NAD, a carrier molecule, for later use in the third stage of energy production.) In the seventh step (Fig. 3.2g), this phosphate, which is just as unstable on the sugar fragment as in ATP, is used to make a new ATP. At this point, the cell has just broken even: two ATP molecules were used in steps one and three, and step seven just restored two ATP molecules, one from each of the halves of the branched pathway.

The last three steps involve some chemical gymnastics that allow the remaining phosphate handles to be replaced onto ATP. The phosphate is first moved to the center carbon atom, poising it (Fig. 3.2h). Then a water molecule is removed, leaving the molecule in an uncomfortable strained conformation (Fig. 3.2i). To get out of this position, the phosphate is allowed to leave in the last step. It is not allowed to simply fall off, however; the energy gained by removing this strain is enough to force the phosphate onto an ATP molecule (Fig. 3.2j).

Glycolysis ends with two pyruvate molecules, each about half the size of glucose, and a net production of two ATP molecules. In most cells, pyruvate is immediately converted to acetyl-CoA by removing a carbon dioxide molecule and attaching the rest to the carrier. The conversion is performed by an enzyme complex, pyruvate dehydrogenase, that is one of the largest enzymes made by modern cells (illustrated in Fig. 1.4 and Fig. 5.5).

If oxygen is not available, the two ATP molecules from glycolysis may be the only source of energy. Yeast in fermenting grape juice and our muscles during rapid, anaerobic exercise have this in common: both are using glycolysis as the major source of ATP energy. In both, there is not enough oxygen to perform the final stage of energy production. The yeast are sealed in a bottle, and our muscles are working faster than our blood can bring in oxygen. In the process, large quantities of pyruvate and hydrogen atoms on NAD are produced. To keep them from building up to toxic levels, a few extra reactions are performed at the end of the pathway. Two enzymes in yeast combine pyruvate and hydrogen to form ethanol (drinking alcohol) and carbon dioxide (the nice bubbles in champagne), both of which are excreted into the medium. An enzyme in our muscles converts pyruvate and hydrogen into lactic acid, which causes the familiar burning pain of overexertion. This is only a short-term solution in our bodies; when we slow down and oxygen is again available, lactic acid is converted back to pyruvate, which then feeds normally into the third stage.

Most of the ATP used by cells is produced in the third stage of energy production, where the carbon and hydrogen atoms in acetyl-CoA are completely broken apart. It starts in the *citric acid cycle* (also known as the Krebs cycle after its discoverer), where the two carbon atoms are combined with oxygen to form carbon dioxide and all of the hydrogen atoms are transferred to carrier molecules: NAD and the related FAD. As in glycolysis, the changes are made in many small steps under the control of specific enzymes. Three molecules of ATP are formed in the three most favorable of these steps.

The largest quantities of ATP are then formed in the process of *oxidative phosphorylation*, where the hydrogen atoms on NAD and FAD are added to oxygen to form water. The details of oxidative phosphorylation have been hotly debated up until the last few decades. There have been two conflicting schools of thought. The traditional camp first postulated that the process occurs much like the reactions of glycolysis. The chemical reaction of adding hydrogen to oxygen was thought to be linked to the formation of ATP by an enzyme, using a few intermediate carriers to perform the task. Unfortunately, these intermediates were never found. Instead, a series of mysterious facts emerged. Surprisingly, the process requires a membrane, which must separate two definite spaces, an inside and an outside. Also, the combining of hydrogen and oxygen could actually be separated from the formation of ATP. These two processes were not occurring simultaneously at the same site, as expected for a glycolysis-like process. The rival camp proposed a radical mechanism explaining these puzzling discrepancies that had more in common with a battery than with any of the other processes then known in the cell. This mechanism has since been widely accepted and has been observed in several other cellular processes.

Oxidative phosphorylation occurs in two separate steps, which are linked electrically. First, the electrons of the hydrogen atoms are removed, transferred down a chain of proteins, and added to oxygen atoms. These oxygen atoms, with extra electrons, quickly grab hydrogen ions to form water (hydrogen ions are common in water and are produced when the electron is removed from a hydrogen atom). At three places along this chain, the transfer of electrons fuels a pump, which pumps positively charged hydrogen ions across the membrane. As more and more electrons flow down the protein chain to oxygen, positive ions accumulate on one side of the membrane, leaving the other side with a negative charge, just like a battery. A charged battery has one pole enriched with negative ions and the other with positive. The separation of charge can be used to do work. We can turn a motor or light a lamp by letting the electrons in a battery flow back, evening out the charge difference between the poles. Cells allow the ions to flow back across the charged membrane through a protein that forces them to make ATP in the process (Fig. 3.3). For each hydrogen atom that combines with oxygen to produce water, the cell can make an incredible three ATP molecules.

The complete oxidation of a single glucose molecule to water and carbon dioxide, from glycolysis to citric acid cycle to oxidative phosphorylation, yields

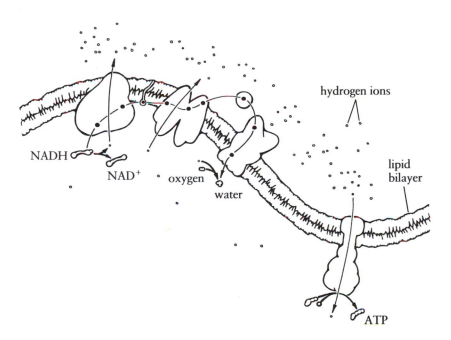

Figure 3.3   Oxidative Phosphorylation

The hydrogen atoms stored on NAD are used to make ATP in the process of oxidative phosphorylation. Electrons are taken from NADH and transferred down a chain of carriers (black dots), ultimately ending up on an oxygen atom in water. In the process, hydrogen ions are pumped across the lipid bilayer, seen here in cross section. Hydrogen ions flow back through the protein complex at right, making ATP. ($1,000,000 \times$)

a total of thirty-six ATP molecules. In comparison, organisms that live without oxygen, using only glycolysis, produce only two ATP molecules per glucose molecule.

Of course there is no manna from heaven. The sugar fueling the production of ATP must ultimately come from somewhere. Green plants provide the wellspring of modern life. In the process of *photosynthesis*, green plants capture energy from the sun and use it to transform inorganic compounds of carbon into biologically usable organic foodstuffs.

Photosynthesis begins with the *light reactions*, which convert the electromagnetic energy of light into the chemical energy of ATP. As in the reactions of oxidative phosphorylation, the light reactions link an electron transfer process to the charging of a membrane. In green plants, flattened, membrane-enclosed sacs (thylakoid disks) are the batteries (Fig. 3.4). Large protein photosystems, embedded in the membranes, capture the energy of light and use it to strip electrons from water, forming oxygen gas. The electrons are then transferred down a chain of enzymes just as in oxidative phosphorylation, and the membrane is charged with hydrogen ions similarly. The charged membrane then is

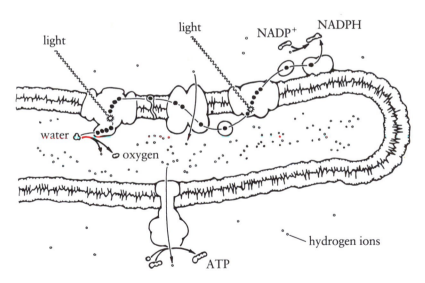

Figure 3.4   Photosynthesis

The energy of sunlight is captured in the light reactions of photosynthesis, creating ATP. Electrons are transferred from water, down a chain of carriers (black dots), to NADP. In the process, hydrogen ions are pumped into the thylakoid disks, seen here in cross section. The concentrated ions are then used to fuel the production of ATP by the protein complex at bottom. (1,000,000 ×)

used, as in oxidative phosphorylation, to make ATP. After a day's work in the sun, the plant has built up a supply of ATP.

In the *dark reactions*, the ATP formed in the light is put to work. Surrounding the thylakoid disks are enzymes that progressively strip oxygen from carbon dioxide and add the carbon to a sugar molecule, fueled by the breakdown of ATP. Ribulose bisphosphate carboxylase, the most plentiful enzyme on earth, forms the key bond between carbon dioxide and the growing sugar chain. This bond is the bond which nourishes us all: a new organic carbon–carbon bond, which, in this or in derivative forms, will be consumed and broken by our cells for energy.

MAKING AN IDENTICAL COPY OF ONESELF

Cells are in a constant state of internal maintenance and repair. While carrying out the tasks of living—finding and digesting food, fighting with competitors, fleeing from predators—cells deftly replace their molecular components that are worn out. They cannot be taken to a shop for repairs, like a broken clock; they must make replacements *in situ*, without ever disturbing the ongoing processes of life.

Cells must also be able to make all of their component molecular machines using only resources available in their environment. Imagine the magnitude of this accomplishment. A simple bacterium contains over one thousand different proteins alone. It makes about five hundred RNA molecules with different orderings of nucleotides, as well as a heterogeneous mixture of lipids and sugar polymers. All of these must be created from the molecules the cell eats, drinks, and breathes.

The ability to create all of their own component parts, coupled with the ability to internally insert newly made components into their own bodies, endows cells with the remarkable ability to reproduce. When environmental conditions are particularly favorable, cells make a big investment of resources and duplicate all of their component parts. The greatly enlarged cells then break in two, and the resulting daughter cells continue a life just like that of their parent.

The key molecular process that makes modern life possible is *protein synthesis*, since proteins are used in nearly every aspect of living. The synthesis of proteins requires a tightly integrated sequence of reactions, most of which are themselves performed by proteins. (Thus posing one of the unanswered riddles of biochemistry: which came first, proteins or protein synthesis? If proteins are needed to make proteins, how did the whole thing get started?)

Each different protein is made according to a blueprint. The unique linear sequence of amino acids in a protein is encoded in the linear sequence of nucleotides in DNA. Because DNA is composed of only four types of nucleotides, compared to the twenty types of amino acids in protein, there cannot be a one-to-one correspondence of amino acid to nucleotide. Cells resolve this problem with the most conservative possible coding: a triplet of nucleotides, three in a row, is used to specify one amino acid. Each position in the triplet can be occupied by one of the four types of nucleotide, so each triplet could potentially specify up to sixty-four amino acids. This is more than enough to specify the twenty amino acids actually used by cells, along with some special triplet codes for starting and stopping. Proteins are built by reading the sequence of nucleotide triplets in DNA and using the information to link amino acids in the proper order.

The cell uses a more circuitous route than might be expected. We could envision a molecular machine that reads DNA directly, walking down a DNA strand and adding amino acids to a growing protein. Perhaps due to the conditions under which life developed (the elusive answer to the riddle posed above), this one-step path is not used by cells today. Instead, cells build proteins in two steps, using an intermediary messenger molecule between DNA and a new protein (Fig. 3.5).

In the first step, *transcription*, the messenger molecule is made according to the information stored in DNA. The enzyme RNA polymerase unrolls a section

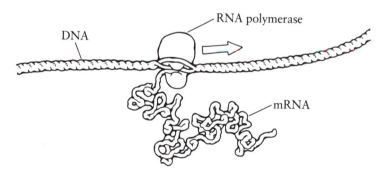

Figure 3.5   Protein Synthesis

Proteins are made in two steps: first the information in DNA is transcribed into mRNA, a messenger molecule, by RNA polymerase, as shown on this page. The information in mRNA is then translated into a sequence of amino acids in a new protein by the combined effort of over fifty molecular machines, as shown on the opposite page. $(1,000,000 \times)$

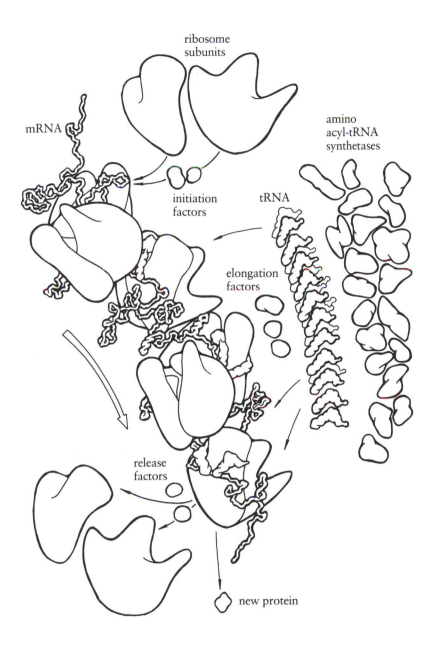

ribosome
subunits

mRNA

amino
acyl-tRNA
synthetases

initiation
factors

tRNA

elongation
factors

release
factors

new protein

of the DNA double helix and, at a rate of about thirty nucleotides per second, builds a strand of RNA complementary to it. When finished, the DNA winds back to its stable, double-helical form. The strand of RNA, known as mRNA ("messenger" RNA), contains exactly the same information as the segment of DNA copied, still in a sequence of nucleotides. But it is a throw-away molecule, to be used and then discarded.

In the second step, *translation*, the sequence of nucleotides in mRNA is read and used to link amino acids in the proper order to form a new protein. Translation requires the combined efforts of over fifty different molecular machines. The actual physical matching of each nucleotide triplet with its proper amino acid is performed by another type of RNA, known as tRNA ("transfer" RNA). Transfer RNA is made in twenty varieties, one for each amino acid. They are L-shaped, with the proper triplet of nucleotides at one end and the amino acid attached to the other end (see Color Plate 2). A separate set of twenty different enzymes (amino-acyl tRNA synthetases) load the proper amino acid onto each type of tRNA.

Proteins are physically built by ribosomes, the engines of protein synthesis. Chaperoned by proteins that initiate and terminate the process, and other proteins that inject the energy for each step, ribosomes walk down a strand of mRNA, align tRNA adapters alongside, and link up the amino acids they carry. At a rate of about twenty amino acids per second, an average protein takes about twenty seconds to build. Over fifty individual protein chains and three long RNA chains combine to form these large molecular factories. That ribosomes are composed of both RNA and protein is provocative—perhaps a relic from the earliest cells. Ribosomes perform the central task of life, so they have probably remained essentially unchanged over the billions of years of evolution.

Apart from proteins, cells must also be able to make all of their other component molecules, using only the molecules in their food. Many bacterial cells can make everything they need starting with only the simple gases available in the air—nitrogen, oxygen, and carbon dioxide—plus a few minerals. A diverse collection of enzymes performs all of the necessary reactions. Some use the energy in sunlight to create sugar from carbon dioxide and water. The sugar then serves as the starting point for other molecules and as a source of energy. Other enzymes trap nitrogen and combine it with hydrogen to form ammonia, which is needed to make amino acids and nucleotides. Other enzymes make lipids and specialized sugars, which are linked together by additional enzymes to form the cell wall. Enzymes link bases, sugars, and phosphates to form nucleotides, which are linked together by a specialized complex of enzymes in the delicate task of duplicating DNA, creating two copies of the blueprint when the cell reproduces.

Gallery of Molecular Structures

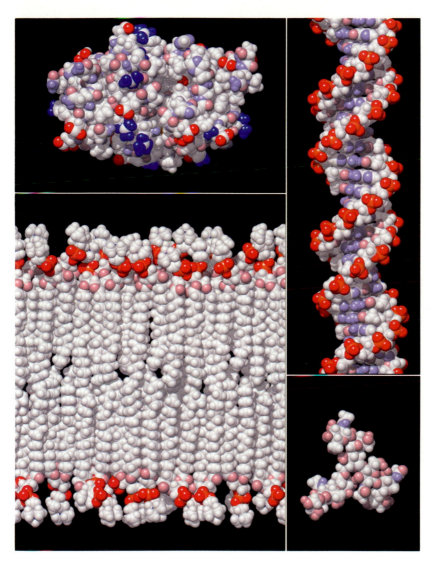

1 Protein, Nucleic Acid, Lipid, and Polysaccharide

The four molecular plans used by living cells have distinctive properties due to their different arrangements of hydrophobic and hydrophilic atoms. In this and the following color figures, hydrophobic carbon atoms are white, mildly hydrophilic atoms are pastel (light blue for nitrogen and pink for oxygen), and strongly hydrophilic atoms carrying a full electrical charge are brightly colored (blue for nitrogen and red for oxygen). Hydrogen atoms are colored according to the atom to which they are bonded and the occasional sulfur and phosphate atoms are colored yellow.

Clockwise from upper left in this figure are: the plant enzyme dihydrofolate reductase, dotted with positively and negatively charged groups; a double strand of DNA (a nucleic acid), over-whelmingly negative in charge; a polysaccharide, covered with slightly hydrophobic groups; and a cross section through a lipid bilayer, almost entirely hydrophobic. (10,000,000 ×)

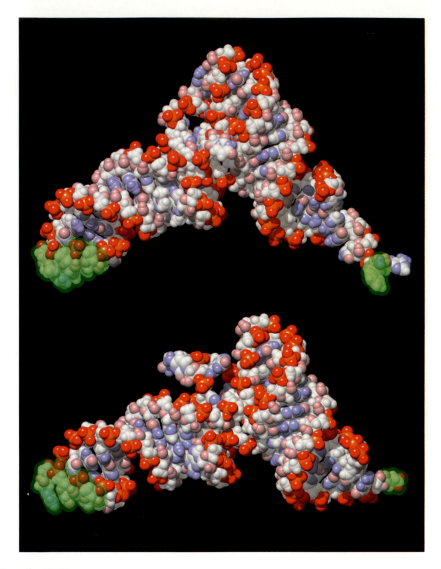

2   Transfer RNA

Transfer RNA is the translator between the codes of nucleic acid and protein—the encryption key. Transfer RNA molecules are L-shaped, twisted strands of RNA. The three bases that read the DNA are at the looped end (highlighted in green at the left side of each), where the chain folds back on itself and several bases are left unpaired. The amino acid, which will be added to a growing protein, is attached at the other end, linked to the last nucleotide in the chain (highlighted in green at the right side of each). The upper tRNA molecule carries the amino acid phenylalanine and the lower tRNA molecule carries aspartate. It is thought that the differences in the region of the "elbow" allow the specific enzyme to recognize the tRNA and load on the proper amino acid. (10,000,000 ×)

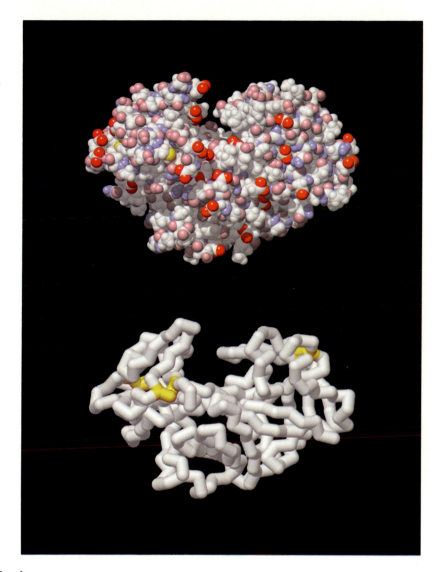

3   Pepsin

Digestive enzymes are generally small and very stable, because they frequently perform in hostile environments. Heat, concentrated salt, or acid will unfold and inactivate a protein. One method of strengthening proteins that act outside the cell is to form several crosslinks between parts of the protein chain, gluing the chain in the right conformation. Pepsin, an enzyme that breaks down proteins in our stomach, has three crosslinks (yellow) between sulfur atoms from different parts of the chain. (In the lower figure, only the positions of each amino acid are drawn, connected by tubes.)

   Our hair is strengthened by similar crosslinks. Hair is composed of long strands of protein lying side by side. Many links between adjacent strands keep the strands from sliding past each other. When you get a permanent wave, the hairdresser applies a chemical that breaks the crosslinks. The limp hair is then placed in the desired position and the crosslinks are reformed. Because this process involves some chemistry with sulfur, it is rather foul-smelling. (10,000,000 ×)

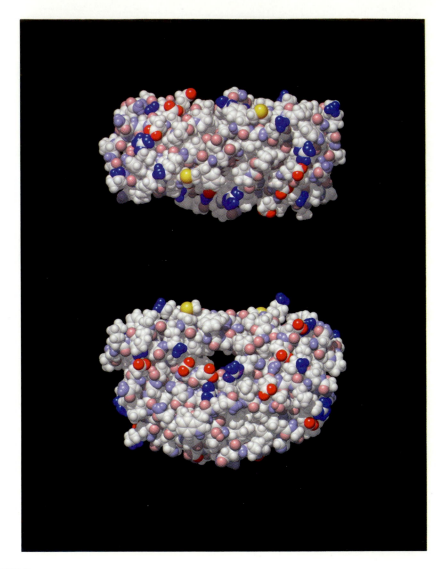

4 HIV Protease

This small enzyme has received a great deal of attention, because it is the first molecule from human immunodeficiency virus to be studied in atomic detail. It is a very specific protease: whereas pepsin will cut proteins in many places, fragmenting them, HIV protease makes only a few specific cuts necessary to the maturation of the virus. (The need for a protease in the viral life cycle is described in Chapter 9.)

Researchers are currently attempting to use the knowledge of HIV protease structure to design drugs combating AIDS, using techniques of computer-aided drug design. They search for new drugs to bind and inactivate this enzyme, blocking a crucial step in the life cycle of the virus. The drugs are designed and screened in the computer. The best candidates are then made by organic chemists, to be tested on active viruses. (10,000,000 ×)

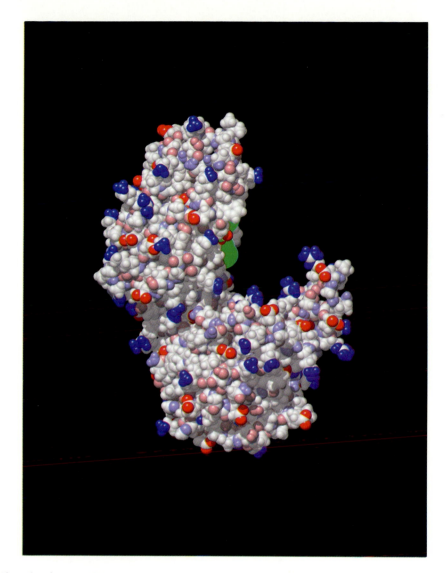

5 Phosphoglycerate Kinase

Phosphoglycerate kinase performs the seventh reaction in glycolysis, described in Chapter 3. It binds to phosphoglycerate (a fragment of glucose) and transfers the phosphate to form ATP. There is a potential problem with this process: if the enzyme were instead to bind phosphoglycerate and a *water* molecule, the phosphate could easily be transferred to water, ruining the chance to make ATP. The enzyme would happily chase around, destroying all of the phosphoglycerate formed in the first six steps of glycolysis, before any ATP could be made.

The unusual shape of phosphoglycerate kinase solves this problem. It is composed of two domains connected by a flexible hinge. Some of the amino acids needed for the reaction are in its upper half, some in the lower half. When the enzyme binds to phosphoglycerate and ADP, they cause the enzyme to close around them. Only then are all of the proper amino acids brought into position, and inside, sheltered from water by the enzyme, the reaction is catalyzed. The trick is that the enzyme must be closed to function, and water and phosphoglycerate alone cannot trigger it to close. (10,000,000 ×)

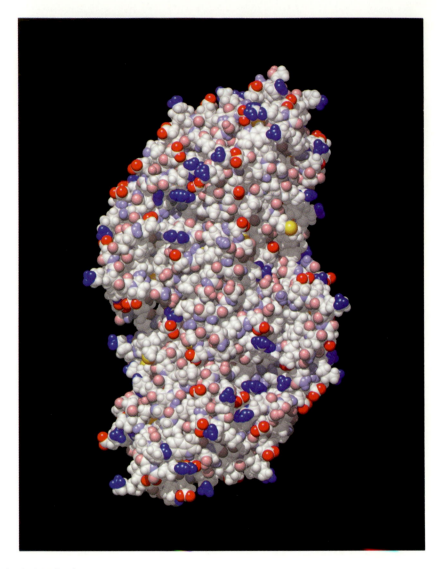

6 Alcohol Dehydrogenase

Alcohol is detoxified in our liver by alcohol dehydrogenase. The amount of this enzyme in our liver is enough to break down about 10 milliliters—a glass of wine or a shot of whiskey—each hour.

Alcohol is a very small, hydrophobic molecule, giving it nearly free reign in our bodies. It passes readily through lipid bilayers and so is able to enter all cells. It passes readily through the barrier separating the blood from the brain—normally a difficult feat—so we feel its effects quickly after drinking. It passes freely to a developing fetus: growing babies drink just as much as their mothers. (10,000,000 ×)

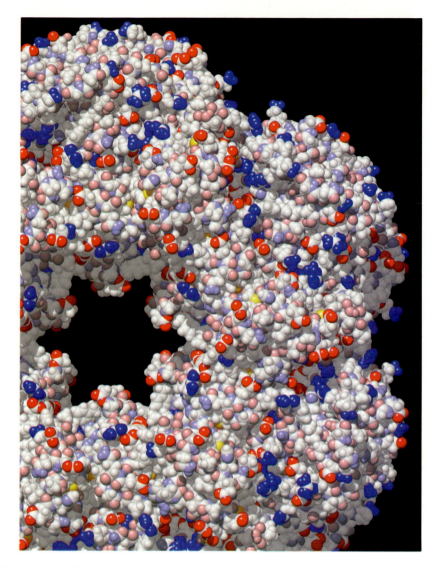

7 Glutamine Synthetase

Glutamine synthetase is an example of a large enzyme complex, formed of twelve identical proteins which pack into a pleasingly symmetric form. Here, we look down a nearly perfect sixfold rotation axis, with proteins arranged in a hexagon. Proteins have been discovered with almost every symmetric arrangement imaginable: threefold axes, fivefold axes, tetrahedrons, and icosahedral polygons have all been observed.

What advantages do large complexes have over single proteins? The most common advantage is regulatory. Many protein complexes, including glutamine synthetase and the proteins in the next two Color Plates, rely on the interactions between their different subunits to regulate their function. Other complexes bring several enzymes together in one functional unit. For example, pyruvate dehydrogenase complex, which takes the product from glycolysis and feeds it into the citric acid cycle, performs several reactions in a row, thus increasing efficiency. (10,000,000 ×)

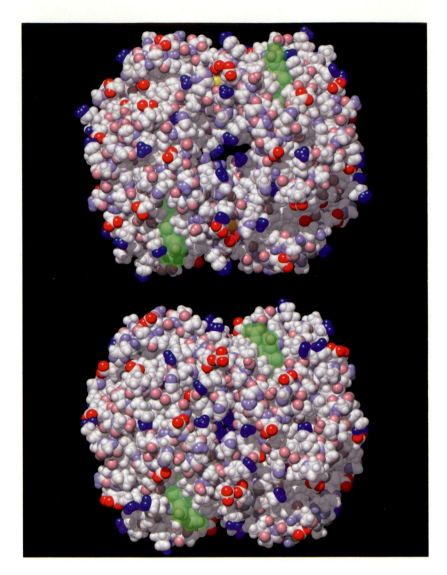

8 Hemoglobin

Red blood cells filled with hemoglobin carry oxygen from our lungs to our cells (see also Fig. 6.6). Hemoglobin is composed of four similar subunits, each of which contains one iron atom held in a flat, square heme molecule (a poisoned heme is pictured in Fig. 10.1). The iron–heme complex gives hemoglobin, and blood, its bright red color.

Hemoglobin is *allosteric* ("other shape"). Changes in shape are important to its function, modulating the affinity of its four binding sites for oxygen. When the first molecule of oxygen binds, hemoglobin changes shape slightly (from the upper shape in the illustration to the lower), making the other three sites more attractive to oxygen. The first oxygen is difficult to add, the other three progressively easier. In our lungs, this first oxygen is forced on because oxygen is plentiful, and the other sites fill rapidly. The opposite is true in our muscles: oxygen is scarce, and the oxygen falls off. Because the fourth oxygen to fall off is difficult to replace, it stays off. ($10,000,000 \times$)

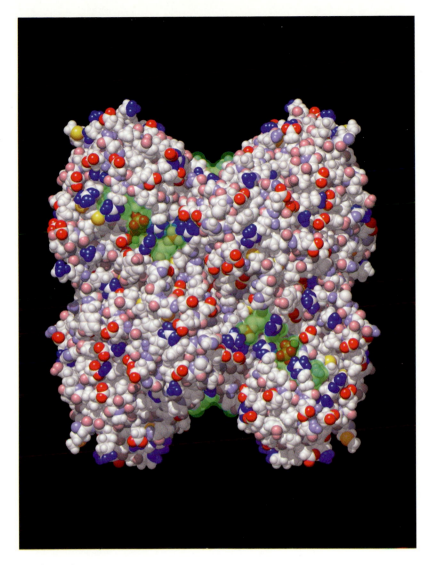

9 Phosphofructokinase

Metabolism is carefully regulated in the cell: food is broken down only when energy is needed. The second step of glycolysis, performed by phosphofructokinase, is a major site of control. Once this step is performed, the remaining steps occur unhindered. Phosphofructokinase is an allosteric enzyme that changes shape in response to the energy needs of the cell. ATP binds at two different positions on the enzyme: in the active sites (two are shown in green on the front face) and in regulatory sites (in green in the clefts at top and bottom). When ATP binds in the regulatory sites, the four subunits of the enzyme are forced to move slightly (to the "other shape") and the active site no longer performs the reaction.

When the cell has a good supply of ATP, there is plenty to bind in the regulatory site, and the enzyme is shut off. When ATP becomes scarce, however, there is no longer enough to fill the regulatory site. The enzyme becomes active and glycolysis again occurs. Thus, glycolysis is self-regulating: the amount of sugar broken down is controlled by the level of the product, ATP. (10,000,000 ×)

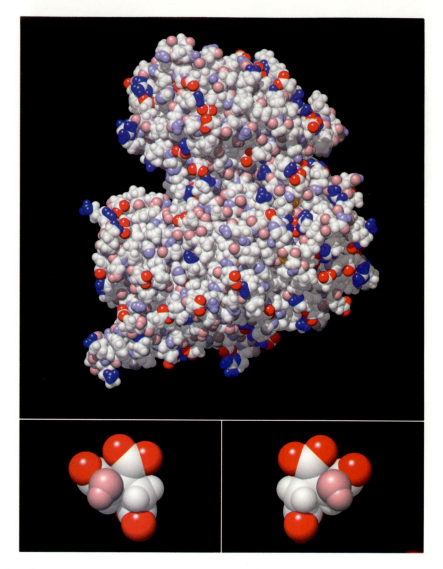

10   Aconitase

Aconitase, an enzyme of the citric acid cycle, can distinguish between the two similar molecules magnified at the base of the illustration. These small molecules are identical except that they are mirror images of one another. Aconitase will accept only the left-hand molecule in the reaction it performs.

Aconitase accomplishes this feat by completely enclosing its substrate molecule. The head at the top of the molecule is connected to the larger body by a long, flexible tether (not seen in the illustration). The protein enfolds its substrate, fitting around it like a glove around a hand. Specific amino acids read each of the projecting groups of the molecule. (top, 10,000,000 ×; bottom, 30,000,000 ×)

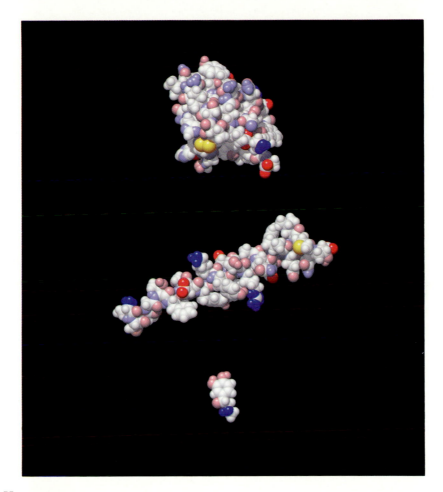

11 Hormones

Most of our cells use glucose as a major source of energy. However, glucose is stored in significant amounts in only two places: the liver and the muscles. Other cells get their supply from glucose circulating in the blood. Several hormones regulate the amount of glucose in the blood. It is important that there be the right amount: too little, and the cells die; too much, and the long-term complications of diabetic hyperglycemia result.

Glucagon and insulin are the two small protein hormones that regulate the normal level of glucose in the blood. They are secreted by cells in the pancreas that constantly monitor the levels of glucose. When the glucose level falls, glucagon (center) is released. Glucagon binds to a receptor on liver cells and initiates a cascade of responses that ultimately releases stored glucose into the blood. When the glucose level is too high—for example, immediately after a meal—insulin (top) is released by the pancreas. Insulin binds to fat cells and stimulates them to take up glucose, reducing the level in the blood. Through the opposite actions of these two hormones, the level of glucose in the blood is held at a constant level.

One of the effects of epinephrine (adrenaline, bottom) is similar to that of glucagon. When we are in danger, epinephrine is released into the blood from the adrenal gland. It affects many parts of the body, making us ready to fight or run away by increasing blood pressure and heart rate and opening the respiratory tract. Epinephrine also signals both the liver and the muscles to release glucose into the blood, giving our body an immediate source of energy, if needed. (10,000,000 ×)

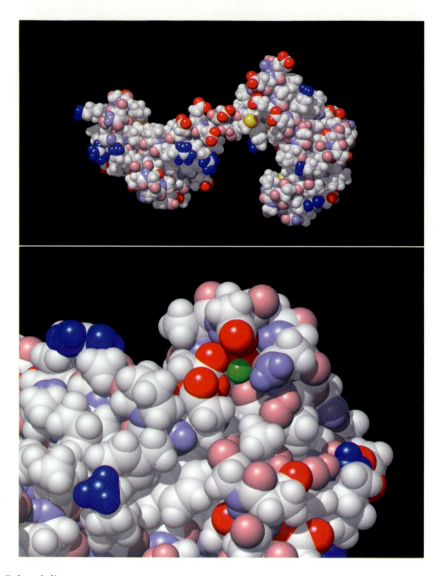

12 Calmodulin

Calmodulin is a small protein that senses the levels of calcium in animal cells. Calcium is used as a signal in many places in our bodies. Calcium is released in our blood when our energy supply is low, and calcium is used locally in nerve transmission and muscle contraction. Minute amounts of calcium are used to transmit these messages, so the proteins that receive them must bind tightly to single calcium ions.

 Many calcium-binding proteins are dumbbell-shaped, with four sites for calcium binding, two at each end. A close-up of one binding site is shown at bottom. Notice the cluster of negatively charged groups, which hold the calcium ion (shown in green) from all sides in an octahedral arrangement. (top, 10,000,000 ×; bottom, 30,000,000 ×)

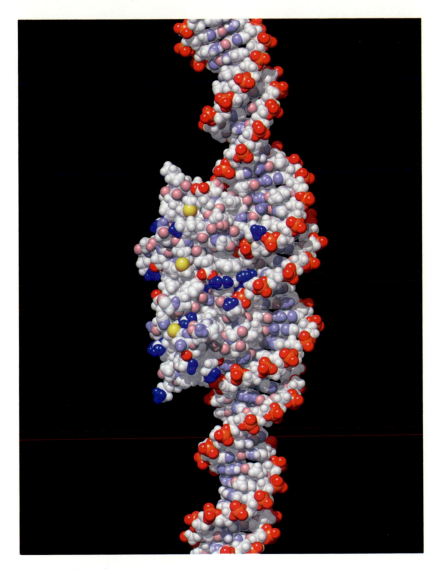

13 DNA-Binding Proteins

The synthesis of proteins is controlled in part by repressor proteins that bind directly to the DNA double helix, physically blocking the polymerase that makes RNA. They recognize a specific sequence of DNA, generally blocking a region of 10 to 20 base pairs.

Repressors form a molecular switch, turning on the synthesis of a given protein only when it is needed. The repressor senses the level of a nutrient in the cell. If the level is low, the repressor remains bound to the DNA, blocking the synthesis of the proteins that act on the nutrient. If the cell stumbles upon a supply of a nutrient, however, the nutrient induces a change in the shape of the repressor that causes it to drop off the DNA. Messenger RNA is then made, which is translated to the protein that will utilize the nutrient. (10,000,000 ×)

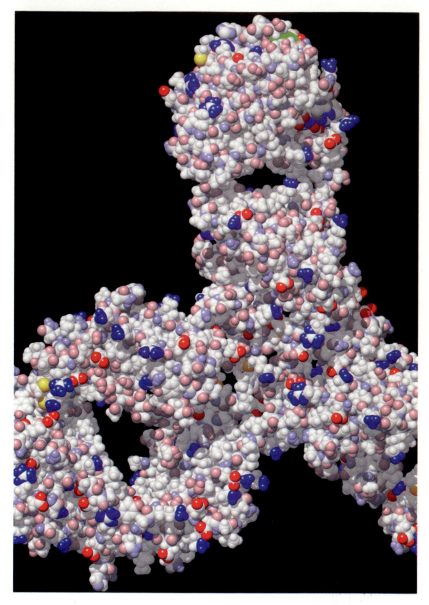

14 Antibody

Antibodies play a central role in our immune system by sensing the presence of foreign substances in our bodies. Antibodies circulate in our blood and bind to unwanted molecules, such as viral or bacterial molecules. Cells in our blood then engulf and digest the antibody-tagged invaders.

Antibodies are Y-shaped proteins with two identical sites for binding foreign molecules at the two tips of the arms of the Y (in green at top; the other tip is off the right edge). The whole molecule is flexible; the arms may flex to allow the antibody to find adjacent sites on an invading cell surface.

Antibodies are exquisitely sensitive to details of structure. Our bodies produce an estimated million different antibodies, all similar in shape, but each able to bind a different molecule. (10,000,000 ×)

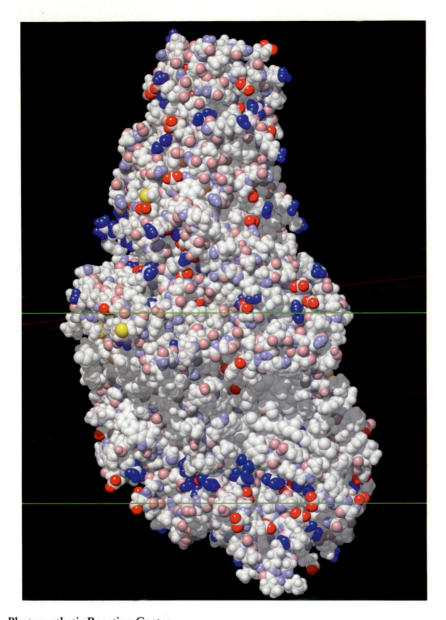

15 Photosynthetic Reaction Center

Modern life on earth is made possible by protein complexes similar to this photosynthetic reaction center. Photosystems are found embedded in membranes in plants and photosynthetic bacteria (green lines enclose the membrane-spanning portion). By converting the energy of sunlight into a chemically useful form, they provide the primary source of energy fueling modern life. Plant cells capture energy from the sun to feed and build themselves. Nearly all other organisms use this energy for their own activity and growth, either by directly eating plants or as part of a web of predation that is ultimately rooted in plant-eaters. (10,000,000 ×)

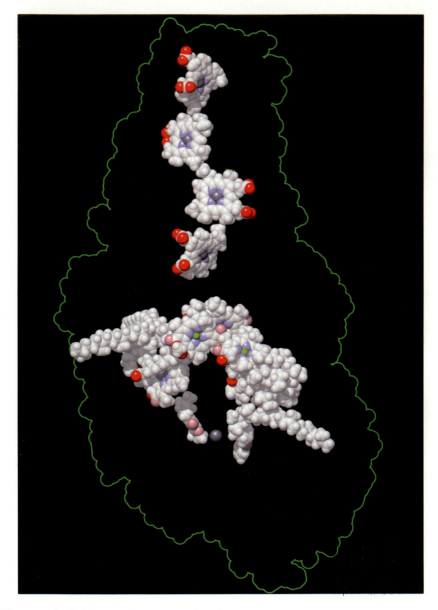

16   Photosynthetic Reaction Center, continued

A chain of prosthetic molecules, including hemes at top, with iron in gray, and chlorophylls at center, with magnesium in green, use the energy of a photon of light to move hydrogen ions across a membrane, from bottom to top in this illustration. The charged membrane is then used by other enzymes to make ATP. (10,000,000 ×)

The cells in our bodies are not quite so self-sufficient. In the course of evolution, we have lost the ability to make many of the molecules needed for our growth, because we can easily get them directly in our diet. We cannot use energy from the sun, so we must eat sugars and fats for energy. To make proteins, we require a few amino acids that we cannot make from scratch. We also require *vitamins* — small molecules that we cannot make, but that are absolutely required by some of our enzymes (vitamins are discussed in Chapter 8). We do, however, still make hundreds of different molecules: lipids, most amino acids, nucleotides, and sugars. Our cells also make many novel molecules: hormones, nerve signal transmitters, painkillers, oxygen carriers, and colorants.

PROTECTION AND PERCEPTION

When interacting with the world, organisms must protect and isolate themselves, but at the same time, they must sense and respond to changing conditions. To perform these opposing functions, modern organisms have developed a bewildering variety of different molecular machines. The molecular machines of synthesis and energy production, discussed above, are nearly identical in all living organisms; the differences between organisms appear mainly in this third class of function. Organisms have evolved many different ways of dealing with their diverse environments. The mechanisms by which they insulate themselves from salty seawater, desert heat, or the thin air of mountains have evolved into a wide range of unique structures. Novel combinations of molecules allow an organism to swim after prey, to run away from a predator, or to taste really bad.

There is one common motif: all organisms use a lipid bilayer as their primary separation between "self" and "outside." Lipid bilayers are resistant to the passage of the large molecules in cells. But a perfectly resistant skin would be useless to a cell. How could food ever get inside? So cells build a diverse set of protein pumps that span the membrane and carry specific molecules from one side to the other, usually requiring the input of some energy. Specific pumps are built to pump amino acids in, to pump urea out, or to trade sodium for potassium. Cells thus carefully regulate the molecules brought inside, transporting nutrients in and waste products out.

Lipid bilayers are tenuous structures, however, and must be buttressed to form a strong protective wall against weather and predators. Here, the diversity of molecular solutions developed to solve a problem appears. The plant cells described in Chapter 7 build a tough sheet of cellulose (a polysaccharide) outside the lipid bilayer. This coat is so tough that it lasts long past the death of the plant. We use the cellulose in wood to build houses and the cellulose in paper to print our words. The cells in our bodies, illustrated in Chapter 6, use an internal scaffolding made of protein, attached to points inside the lipid bilayer and linked to a mesh of protein fibers that run throughout the cell. The bacterium described in the next chapter builds two concentric lipid bilayers, with a tough protein–sugar sheath tethered in between.

Diversity is the rule when we look to the molecular machinery of perception and reaction. We might expect this, as the environment is the single largest influence on the evolution of organisms. They have been specifically selected to best sense, exploit, and multiply in their particular environmental niche. Contrast two evolutionarily distant relatives: the intestinal bacterium *Escherichia coli* and its host, ourselves. We span the spectrum of complexity in living organisms. The bacterium has minimal capability for perceiving and reacting to short-term changes in its environment, whereas the major portion of our body is devoted to these tasks.

Escherichia coli cells commit less than 5% of their molecular machinery to motion and perception, allowing only the simplest of responses. They can be observed swimming through the water in a straight line, under the power of long, corkscrew-shaped flagella rotated by a motor complex embedded in the cell wall. Inside the cell, a few proteins constantly sense the level of nutrients in the surrounding water. If the nutrients are remaining steady or increasing in level, nothing happens. But when the level begins to fall, these proteins interact with the flagellar motor, causing it to temporarily reverse direction. Due to the handedness of the helical flagella, the reverse in direction causes them to flail in all directions, making the bacterium stop and tumble. Soon, the motor returns to normal and the bacterium continues swimming in a new, random direction. Under this simple system, requiring only a motor and a few sensing proteins, the bacterium manages to head toward a source of food, but it does so very inefficiently. If swimming toward food, it does not stop and tumble, but continues straight. If nutrient levels are dropping, however, it picks a new, random direction by tumbling, bumping around in a series of short runs until it finds a better direction. In the environment of our intestines, this almost random approach is apparently adequate.

Our bodies, in contrast, are built for specific, directed motion under the control of detailed perception. The bulk of our body weight is dedicated to sense, reaction, and motion. Cells in our retinas are filled with a protein which senses light, light which is focused by layers of eye lens cells packed full of clear crystallin proteins. Cells in our skin spin enormously long strands of protein, our hair. Other skin cells sense their slightest movement. These and other sensory data are transmitted and processed by specialized cells in our brain that carry electrical currents propagated by proteins and insulated by lipids. Fine control of movement is accomplished by an enormous structure of mineralized bone cells, moved by muscle cells filled with proteins that do nothing but contract, all glued together by connective tissue cells that build tough layers of sugar and protein.

The common thread of life on earth still shows through this diversity, however, tying the simplicity of the bacterium to the complexity of our bodies: all of these unique structures are made of the same four molecular components— proteins, nucleic acids, lipids, and polysaccharides.

II: Molecules into Cells

The organisms that inhabit the
Earth, in all of their diversity,
look remarkably similar at a
magnification of one million
times. In each, molecular
machines of protein, nucleic acid,
lipid, and polysaccharide are
adapted to fit every need.

Chapter 4

Escherichia coli:
One of the Simplest Cells

A bacterium that inhabits our intestinal tract, *Escherichia coli*, has the distinction of being the most biochemically defined organism known to science. It has been central to the study of biochemistry since its discovery in 1885 by Escherich, due in part to its general availability and ease of growth, and in part to plain serendipity. *Escherichia coli* has played a major role in many of the seminal discoveries of biochemistry: the genetic code, glycolysis, and the regulation of protein synthesis. To quote a review of the molecular biology of *Escherichia coli*: "Not everyone is mindful of it, but all cell biologists have two cells of interest: the one they are studying and *Escherichia coli*!"*

Individual *Escherichia coli* cells are small and rod-shaped (Fig. 4.1). Under very rich conditions they grow to about four micrometers in length; under harsh conditions they may be half this size. Through a light microscope, they are seen swimming through their medium with a distinctive "swim-and-tumble" pattern discussed in Chapter 3. With an electron microscope, which allows a magnification just short of that needed to resolve individual molecules, many morphological details may be discerned. Extending from the cell surface are about ten long, screw-shaped flagella, the cell's propellers. The cell body is surrounded by a tough cell wall, composed of several concentric layers. Inside is the cytoplasm, which looks granular in the electron microscope due to the dense packing of ribosomes and proteins. In the center is a less crowded region into which the cell's DNA is packed.

On the following pages, the illustrations show five small square sections of an average *Escherichia coli* cell magnified by one million times. These simulated views reveal the size, shape, and placement of individual molecules as they perform their many tasks in the cell.

---

*Schaechter M, Neidhardt FC in *Escherichia coli and Salmonella typhimurium: Cellular and Molecular Biology*. (Neidhardt FC, editor) Washington, DC: American Society for Microbiology, 1987, p. 2.

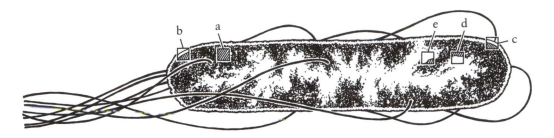

Figure 4.1 *Escherichia coli*

The five boxed regions are enlarged on the following pages. a. Cytoplasm. b. Cell wall. c. Cell wall with flagellum. d. Nuclear region. e. Nuclear region showing DNA replication. (30,000 ×)

CYTOPLASM

Bacterial cytoplasm is a teeming soup of molecules. The cube shown in Figure 4.2, 100 nanometers on a side, contains roughly 450 proteins, 30 ribosomes, 340 tRNA molecules, and several long mRNA molecules. Between these large molecular machines, thousands of small molecules circulate: 30,000 small organic molecules—including amino acids, nucleotides, sugars, ATP, NAD, and a host of others—and 50,000 ions. The space that remains, about 70% of the volume, is filled with water, surrounding and lubricating the collection. (The small molecules are shown only in the right upper corner of Figure 4.2. In the remainder of the book, they are omitted for clarity.)

Two major tasks are under way amidst the bustle. Over half of the molecular machines are actively involved in synthesis of new proteins. Ribosomes, the engines of protein synthesis, dominate the view. Messenger RNA forms a twisted path between the ribosomes, surrounded by the twenty types of tRNA molecules. Of the four hundred or so protein molecules in each 100-nanometer cube, about one-quarter are involved in protein synthesis, loading amino acids onto tRNA and chaperoning the various stages of adding each to a new protein.

The remaining proteins are mainly involved in energy production. About one-quarter are the enzymes of glycolysis, and another quarter are the enzymes of the citric acid cycle. The remaining quarter are a diverse mixture of enzymes, performing a variety of synthetic and degradative tasks. They are typically present in much lower concentrations than the molecules of protein synthesis and energy production. If, however, the cell comes upon a rich source of a particular food molecule, such as lactose, the cell can quickly make a large quantity of the three or four degradative enzymes needed for its use. In an average cell, however, these enzymes would barely be seen.

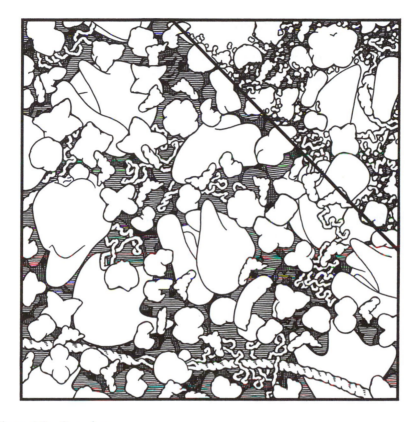

Figure 4.2 Cytoplasm

The cytoplasm of *Escherichia coli* is a complex mixture of proteins, nucleic acids, and small molecules. For clarity, the small molecules are drawn only in the upper right corner; in the remaining illustrations, only proteins, nucleic acids, lipids, and polysaccharides are drawn. At the bottom, a strand of DNA is being transcribed to mRNA, which is immediately translated into new proteins by the closely packed ribosomes. Between the ribosomes, proteins in many shapes and sizes are breaking down small molecules for energy and synthesizing new molecules for growth.

   Of course, all of these molecules are in rapid, chaotic motion, and are frozen in this picture. Remember the crowded airport terminal described in the Introduction. (1,000,000 ×)

CELL WALL

The cell wall of *Escherichia coli* is composed of several layers, successively insulating the interior of the cell from the environment (Fig. 4.3). The outer membrane is a protective barrier, albeit a leaky one. It is a lipid bilayer with specialized lipopolysaccharides making up most of its outer face. As the name

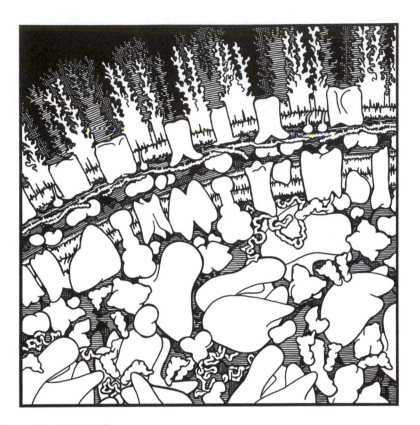

Figure 4.3   Cell Wall

The *Escherichia coli* cell wall, seen in cross section, is composed of several concentric layers. The water outside the cell is at top, in black, and the densely packed cytoplasm is at bottom. The layers of the cell wall from top to bottom (outside to inside) are: the outer membrane, with its gluey polysaccharides extending outward and porin pores spanning the bilayer; the thin layer of crosslinked peptidoglycan strands, connected to the outer membrane by small lipoproteins; the periplasmic space, containing a few small proteins; and the complex inner membrane, studded with many different proteins. (1,000,000 ×)

suggests, lipopolysaccharides are hybrid molecules consisting of lipids linked to several long strings of polysaccharide. The lipid anchors in the membrane, while the polysaccharide chains extend out into the surrounding liquid, forming a gluey, protective coat. The inner face of the membrane is composed of typical lipids.

The outer membrane is not completely impervious, however, as food particles must be able to enter the cell. The membrane is spanned by porin molecules, proteins whose only function is to form a hole through the membrane. Bacterial cells secrete digestive enzymes into the surrounding space that break foodstuffs into small fragments. The holes in porin are large enough for the fragments of food to enter, but small enough to keep the machinery of the cell inside.

Immediately inside the outer membrane is a tough layer of peptidoglycan, a crosslinked sheet of polysaccharide and protein. This meshwork bag envelops the whole cell and holds it in its proper shape. The outer membrane is anchored to the peptidoglycan layer by lipoprotein, another hybrid molecule composed of a small protein with a lipid attached to one amino acid. The lipid inserts into the membrane and the protein binds to the peptidoglycan layer. Lipoprotein acts as an adapter between the hydrophobic lipid membrane and the hydrophilic sugar–protein peptidoglycan.

Inside the peptidoglycan layer is the periplasmic space, just wide enough for a few proteins. Enzymes reside here that continue the digestion of entering food particles. Other proteins collect a specific fragment from this digested material, such as a certain sugar or amino acid, and deliver it to the inner membrane to be transported inside to the cytoplasm. The proteins that sense the scent of food are also found here, where they monitor the level of nutrients and direct the flagellar motors to reverse direction when a tumble is warranted.

The inner layer of the cell wall is the cytoplasmic membrane, or inner membrane. It is less than half lipid; the rest is a diverse collection of proteins embedded in or spanning the membrane. Some are efficient pump proteins, transferring specific molecules into and out of the cell according to current needs. One pump transfers sodium out and potassium in. Another takes delivery of a sugar molecule from a scavenger protein in the periplasm and transfers it inside, requiring energy in the process. A complex pump binds the ribosomes on the cytoplasmic side and carries new periplasmic proteins across the membrane.

In addition to transporting substances into and out of the cell, the inner membrane functions in energy production. Oxidative phosphorylation (described in Chapter 3) requires a membrane separating two compartments. In bacteria these two compartments are the cytoplasm and the periplasmic space. Hydrogen ions are pumped from cytoplasm to periplasm, inside to outside, charging the membrane. Cells then use the favorable process of their return to make ATP or fuel the turning of their flagella.

The ten or so randomly placed flagella are the most impressive features of the cell wall (Fig. 4.4). Each rises from a motor complex which spans the entire cell wall, extending 5 to 10 micrometers into the surrounding water in a gentle helical curve. The motor turns the flagellum in either direction, clock-

Figure 4.4   Flagellum and Flagellar Motor

The ten or so flagella are turned by flagellar motors, propelling *Escherichia coli* cells at a speed of 10–20 micrometers per second. Each motor is a large complex of proteins that spans the entire cell wall, seen here in cross section. The flagellum, at this magnification, would extend 50 to 100 times the width of this illustration. (1,000,000 ×)

wise or counterclockwise. The rotation is fueled by the charged inner membrane, as in photosynthesis and oxidative phosphorylation. The energy gained by allowing about one thousand hydrogen ions to return to the cytoplasm is converted into one rotation of a flagellum.

NUCLEAR REGION

The library of the cell is at its heart. The DNA of *Escherichia coli* is folded into a small nuclear region at the center of the cell (Fig. 4.5). All but a few proteins are encoded in one large circular DNA molecule about half a millimeter in diameter. Bacteria also contain plasmids, small circles of DNA that encode two or three proteins each. Plasmids are easily transferred between individual bacterial cells, and occasionally to other species of bacteria. Plasmids often encode proteins that confer drug resistance. As a result, the obvious advantage—resistance to an antibiotic—is easily shared in a population of bacteria.

The half-millimeter circle of DNA must be packaged into a space less than one-hundredth its diameter at the center of the cell. The protein HU facilitates this process. DNA winds around HU cores, compacting and changing its direc-

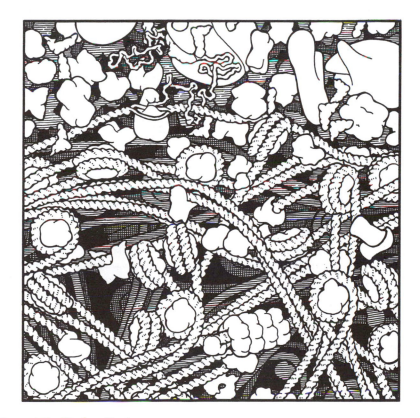

Figure 4.5   Nuclear Region

The nuclear region is filled with a single, large circle of DNA, twisted and folded around HU proteins. The cytoplasm and the nuclear region are not separated by a physical barrier, as in our cells. The nuclear area is relatively free of proteins because the dense packing of DNA strands acts like a strainer, excluding the ribosomes and proteins in the cytoplasm. (1,000,000 ×)

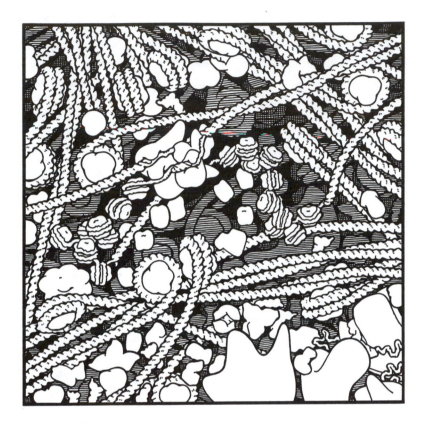

Figure 4.6   DNA Replication

DNA is duplicated by the enzyme complex DNA polymerase, seen at center. The DNA double helix is unwound, and a new strand is built complementary to each original strand. The transiently exposed single strands are protected by small, single-strand-binding proteins. (1,000,000 ×)

tion. These little bundles easily fall apart, so the DNA may be packaged and unpackaged according to need.

The nuclear region is not as densely packed with proteins as is the cytoplasm, but it still holds a diverse set of DNA-binding proteins. These proteins sit on the DNA and block the synthesis of mRNA (see Color Plate 13). Their job is to hold a certain section of the DNA in storage until the protein encoded there is needed. For instance, when the cell comes upon a region rich in the sugar lactose, the lac repressor protein senses the increase and literally falls off the DNA. Messenger RNA is then free to be transcribed from that stretch of DNA, and the enzymes involved in the utilization of lactose are synthesized. Often the entire process of sensing the concentration of a nutrient such as lactose is performed by a single copy of the protein per cell.

Under ideal conditions, with plenty of nutrients and minimal crowding, *Escherichia coli* cells can reproduce in just under thirty minutes. In this time, the cell duplicates everything in its body: ribosomes, proteins, cell wall components, and the rest. The cell pinches in two, forming two identical cells, each of which will be able to reproduce in another half-hour. Central to the process of reproduction is the duplication of the cell's genetic information. The circular DNA strand is duplicated by DNA polymerase. This complex of enzymes pulls apart the double helix and builds a complementary mate to each half, yielding two daughter circles (Fig. 4.6). Replication begins at one site on the circle and proceeds simultaneously in both directions, ending at a site on the opposite side.

DNA replication is very efficient. About eight hundred new nucleotides are added every second, taking about fifty minutes to duplicate the entire circle of 4,720,000 nucleotides. But if *Escherichia coli* cells divide every half hour under favorable conditions, there is not enough time to duplicate the entire store of DNA before each division. *Escherichia coli* resolves this problem by initiating new DNA duplications before the previous round has finished. When a round of duplication is completed, and the two circles fall apart, each new circle is already about halfway through a second round of duplication. Bacterial cells are truly adapted to rapidly exploit their environment!

Chapter 5

Baker's Yeast: The Advantages
of Compartmentation

A breakthrough occurred in the history of cells about 1.5 billion years ago. Simple, bacteria-like organisms had been around at least 2 billion years when a radically new design of the cell body appeared. These new cells were filled with small, separate compartments, each fully surrounded by a watertight lipid bilayer. The advantage of internal compartmentation is profound: the different functions of the cell can be performed in specialized locations. All of the molecular machinery of transcription or of oxidative phosphorylation can be concentrated in one place, allowing more efficient performance and finer control.

Compartmentation signaled a major break in the evolution of cells, forming two distinct family lines. The simple cells, with their molecular machinery jumbled together in a single compartment, are the ancestors of modern bacteria, like *Escherichia coli* of the last chapter. The new, compartmented cells gave rise to all other life, from tiny protozoa to mammals like ourselves to the tallest trees. We now separate modern forms of life into two basic types. The simple, uncompartmented bacteria are *prokaryotes*. The complex, compartmented cells are *eukaryotes*. (Prokaryote for "before kernel" and eukaryote for "true kernel"; the kernel referring to the easily observed central compartment, the nucleus.)

Baker's yeast, *Saccharomyces cerevisiae*, is one of the simplest eukaryotes (Fig. 5.1). Yeast has a rich scientific and economic history. It has been used since prehistoric times for leavening bread and brewing alcoholic beverages. Yeast cells were among the first "animalcules" discovered in the late seventeenth century by Antonie Van Leeuwenhoek using his newly invented microscopes, thus opening up a new world of microscopic life. As the nineteenth century opened and the foundations of biochemistry were being laid, a raging controversy centered on a question about yeast: Was fermentation actually caused by living organisms? The vitalists believed that the answer was "yes." They held that yeast cells absorbed sugar, used it as food, and excreted alcohol and carbon dioxide as waste products. The equally vociferous mechanists held,

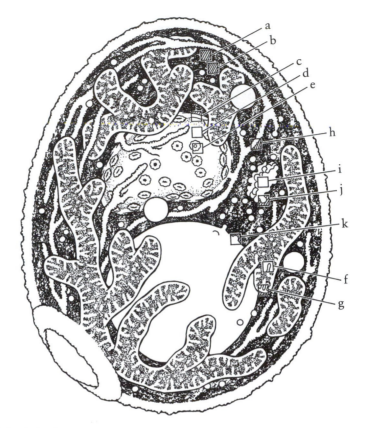

Figure 5.1 Baker's Yeast

The eleven boxed regions in this typical yeast cell are enlarged on the following
pages. a and b. Cytoplasm. c. Nuclear membrane. d. Nuclear matrix. e. Nuclear
pore. f and g. Mitochondrion. h. Endoplasmic reticulum. i. Golgi apparatus.
j. Coated vesicle. k. Vacuole. The crater-shaped structure on the lower left side of
the cell is a bud scar, left when the cell separated from its parent. (25,000 ×)

however, that yeast was merely a chemical substance, not a living entity, which
acted directly on sugar molecules. Pasteur finally resolved the problem in favor
of the vitalists with a series of careful experiments. ("Vitalism" has a different
connotation now: the belief that all organisms have a "life essence" in addition
to their purely physical-chemical bodies.)

This was not the end of controversies centered on yeast. Later, researchers
found that a solution of ground-up yeast, in which the cells had been destroyed,
could still ferment sugar. A chemical substance called "zymase" was postulated
to be the agent of fermentation. Yeast was thought to be merely a cell that

secreted zymase—a new victory for the mechanistic camp. Further experimentation showed, however, that the ground-up yeast solution was unstable compared to fermentation in the cell, losing strength in a short period of time. The observed fermentation was merely the components of the cell playing out their individual roles without the living link to growth and reproduction, like a chicken running around the barnyard after its head has been cut off. The living, vitalistic view of yeast was completely vindicated.

Whereas the prokaryotic bacterial cell is simple in morphology—a "bag of enzymes"—the eukaryotic yeast cell is filled with complexity. Through the electron microscope, we see a large nucleus, filled with the cell's blueprint of DNA and surrounded by a double membrane pierced with beautifully symmetric pore proteins. We see thick, branching mitochondria, bounded by two concentric membranes and filled to the limit with proteins of energy utilization. A large, clear vacuole (called a lysosome in other cells), filled with a watery solution of digestive enzymes, occupies a major portion of the cell. The endoplasmic reticulum, studded with ribosomes making proteins, forms a delicate net of tubules and sheets throughout the cell. If lucky, we see a small Golgi apparatus, a stack of disk-shaped compartments where sugars are added to newly built proteins.

Of course, with the addition of all this new structure, a price must be paid. Since yeast cells do not perform all of their functions in one cytoplasmic compartment, as in bacteria, they must have an elaborate overhead of communication and transport. As new proteins are synthesized, they must be transported to the proper compartment. Also, the cell has become larger and requires a specialized set of scaffolding molecules for structural support.

CYTOPLASM

Yeast cytoplasm is markedly different from the cytoplasm of bacterial cells, as the larger yeast cells require considerable support to maintain their functional shape (Fig. 5.2). One solution, used by most eukaryotic cells, is to build an internal cytoskeleton composed of stiff protein filaments. Yeast cytoplasm is crisscrossed with filaments of three major types. Actin is by far the most common structural element. The cytoplasm is permeated with actin filaments that run from edge to edge. Proteins with descriptive names like filamin and fimbrin crosslink actin filaments, forming a strong mesh of fibers that give the cell its defined but still elastic shape.

Two other types of filament run through the cell, in smaller number than the actin filaments. Intermediate filaments, larger than actin filaments and with a distinctive ribbed appearance, often run perpendicular to the actin

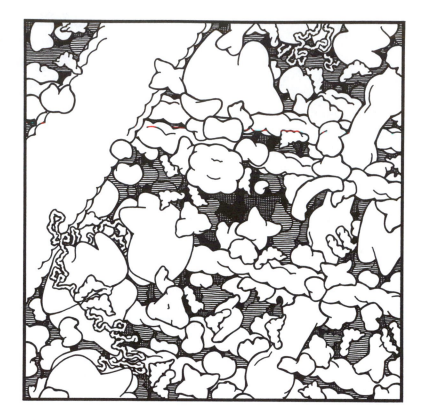

Figure 5.2   Cytoplasm: Cytoskeleton

Yeast cytoplasm is crisscrossed with structural filaments, together forming the cytoskeleton. The largest are microtubules; one runs diagonally at upper left. The smallest, and most plentiful, are actin filaments; several run horizontally in this view. Intermediate filaments, as the name implies, are intermediate in size; one runs diagonally at right, approximately parallel to the microtubule. (1,000,000 ×)

filaments, warp to actin weft. Microtubules are the largest and the least common. They are the railways of the cell: the proteins kinesin and dynein bind to microtubules and, using ATP for energy, walk from one end to another, pulling vesicles along with them.

Many of the cell's synthetic reactions are performed in between the network of cytoskeletal filaments. The machinery of translation, consisting of ribosomes, tRNA and tRNA synthetases, builds proteins from mRNA in the cytoplasm (Fig. 5.3). Unlike prokaryotic mRNA, however, eukaryotic mRNA is not made in the cytoplasm. It is transcribed from DNA in the nucleus and transported into the cytoplasm for translation. The mechanics of yeast protein synthesis are nearly identical to those in bacteria. In the course of evolution, a few changes have been made, most notably, the ribosomes have evolved to a larger size. They still, however, perform exactly the same function: lining tRNA molecules along the mRNA strand and linking up amino acids.

Many other chemical reactions are also underway in the cytoplasm. Glycolysis is performed here, starting the breakdown of sugars. But the later steps of energy production—the citric acid cycle and oxidative phosphorylation—are sequestered in the mitochondria, and many of the earlier steps of digestion occur in the vacuole. Nucleotides and amino acids are synthesized in the cyto-

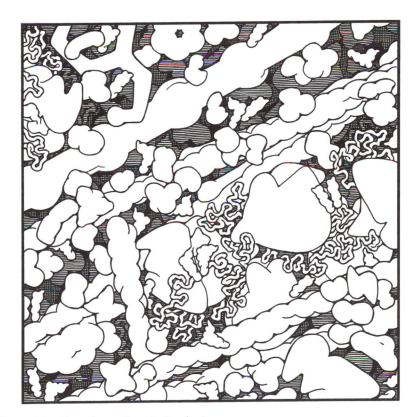

Figure 5.3   Cytoplasm: Protein Synthesis

Many synthetic tasks are underway in the cytoplasm. As in bacteria, the major task is protein synthesis. Notice that the cytoplasmic ribosomes of yeast are larger than those of bacteria, with a distinctive difference in shape. An intermediate filament runs diagonally at upper left. (1,000,000 ×)

plasm. The cytoplasm tends to be the site of reactions that do not fit the specialized tasks of the other compartments—the general housekeeping reactions of the cell.

MITOCHONDRIA

Mitochondria are the power generators of eukaryotic cells. Surprisingly, under the microscope they look much like whole *Escherichia coli* cells. They are bounded by two membranes (Fig. 5.4). The outer membrane, separating

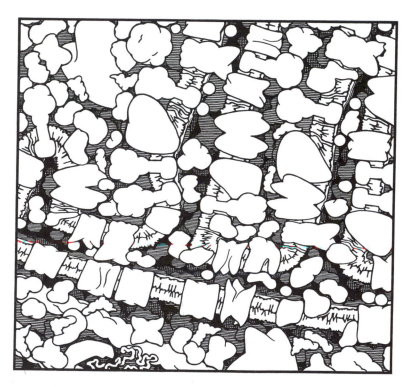

Figure 5.4   Mitochondrial Membranes

Mitochondria are surrounded by two concentric membranes. In this cross section, a sliver of cytoplasm runs across the bottom and the outer membrane of the mitochondrion is immediately above it, pierced by pore-forming proteins. The folded inner membrane of the mitochondrion is above, embedded with energy-producing proteins. A portion of mitochondrial matrix is seen at upper left. (1,000,000 ×)

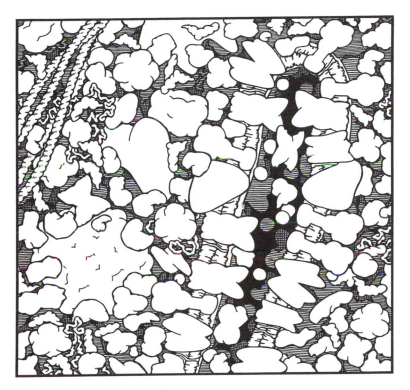

Figure 5.5   Mitochondrial Interior

Mitochondria are filled to their limit with proteins, ribosomes, and nucleic acids.
Notice the presence of DNA, tRNA molecules, and small ribosomes like those of
bacteria. A fold of the inner membrane extends from the bottom. The enormous
spherical protein at lower left is pyruvate dehydrogenase complex, the enzyme that
links glycolysis with the citric acid cycle by converting pyruvate to acetyl-CoA (see
Chapter 3). (1,000,000 ×)

the interior of the mitochondrion from the cytoplasm, is pierced by molecules
reminiscent of bacterial porins. The intermembrane space contains a host of
proteins similar to the periplasmic proteins of *Escherichia coli*. The inner membrane, pleated and folded in mitochondria, is studded with the proteins of
oxidative phosphorylation—the same proteins found embedded in the inner
membrane of the bacterium.

Mitochondria hold further surprises. Along with a highly concentrated collection of enzymes, we find ribosomes, DNA, and tRNA inside mitochondria
(Fig. 5.5). They contain a full set of molecular machinery to build new pro-

teins, separate from that in the cytoplasm. And even more unusual, the ribosomes are smaller than those in the cytoplasm, looking more like bacterial ribosomes!

These remarkable similarities prompted an interesting theory of the origin of mitochondria, which is now widely accepted. It is thought that an ancient bacterium took to living inside another cell several billion years ago, perhaps entering as a parasite or perhaps surviving the process of being eaten. The bacterium proceeded to reproduce inside the cell, always sending copies of itself to each daughter cell when the host divided. Gradually, both the engulfed bacterium and the cellular host became increasingly reliant on one another and slowly divided up the tasks of living. The bacterium focused on generating ATP, the cell on hunting down food and dealing with the environment. This ancient bacterial invader, now more partner than parasite as a mitochondrion, has retained a limited autonomy: modern mitochondria still build a set of protein synthesis machinery separate from that of the nucleus and cytoplasm.

ATP synthesis is the primary process performed in the compartments of the mitochondria. The enzymes of the citric acid cycle are found inside the mitochondria, along with enzymes involved with the breakdown of fats. The inner membrane provides the necessary separation of two spaces needed for oxidative phosphorylation; each mitochondrion is a small cellular battery. This membrane is actually mostly protein. The large enzyme complexes involved in the addition of oxygen to hydrogen are packed so closely that many touch, with only bits of lipid in between.

NUCLEUS

The nucleus is the cell's library, storing the delicate strands of DNA and protecting them from the rigors of the cytoplasm. Transcription of DNA into mRNA is performed here as well. The nucleus specializes in archiving, reading, and regulating the use of the information held in DNA.

The nucleus is surrounded by a double membrane (Fig. 5.6), which is in fact an extension of the endoplasmic reticulum (discussed below). The double membrane is strengthened inside by crisscrossed layers of protein filaments similar to the intermediate filaments found in the cytoplasm. Inside this protective barrier, DNA is stored in several chromosomes ("colored bodies," so-called because they stain with cytochemical dyes). Each chromosome is a linear strand of DNA millions of nucleotides in length. Much of the DNA is wrapped around protein cores, forming nucleosomes similar to the HU protein complexes of

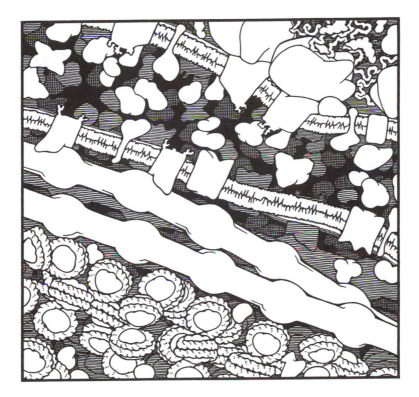

Figure 5.6   Nuclear Membrane

The nucleus is surrounded by a double membrane continuous with the endoplasmic reticulum (compare this view with Fig. 5.9). In this cross section, the cytoplasm is at upper right and the nuclear interior, with DNA coiled into nucleosomes, is at bottom. Intermediate filaments strengthen the inner side of the nuclear membrane. (1,000,000 ×)

bacteria. Nucleosomes protect DNA and aid in compacting the long stands into the small space of the nucleus.

Only transcription (the first step of protein synthesis) occurs in the nucleus (Fig. 5.7). Yeast RNA polymerase, the transcribing enzyme, is more complex than the analogous enzyme of bacteria. In fact, yeast uses three different, larger forms. One form transcribes only mRNA; the others transcribe the RNA chains in ribosomes and tRNA.

There is yet another wrinkle on transcription in eukaryotes: the coding of

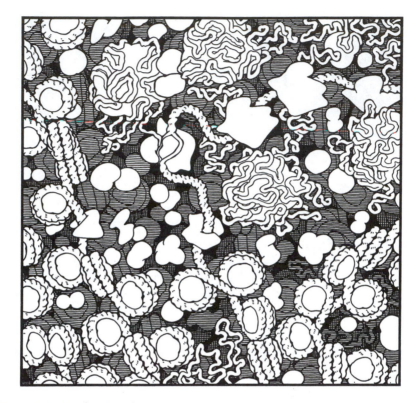

Figure 5.7   Nuclear Interior

The nucleus is filled with DNA, along with proteins involved in its storage, transcription, and regulation. In the center portion of this view, several RNA polymerase molecules are transcribing mRNA from DNA, and the new mRNA is being gathered and processed in large splicosome complexes. The other DNA in this view, not currently in use, is wrapped into protective nucleosomes. (1,000,000 ×)

proteins in eukaryotic DNA is not continuous, as it is in bacteria. Most proteins are encoded in several short passages, separated by lengths of DNA which are not used. When a eukaryotic mRNA is made, therefore, it contains useless stretches that must be edited out to generate the proper message for building an active protein. Splicing is performed in splicosomes, large complexes of proteins that edit the RNA strand as it is formed. It is not known whether these interruptions are used by cells—for regulation or to facilitate the mutation of new enzymes—or whether they are evolutionary debris, left over from billions of years of shuffling information.

In yeast cells, transcription and translation never occur concurrently, as they do in bacterial cells. This is a definite advantage to yeast cells. There is now an extra step at which the process can be regulated: the transport of mRNA from its source in the nucleus to the cytoplasm, where it is translated. After new mRNA molecules are built and edited, they are transported to the edge of the nucleus, protected in a large complex of proteins. At the nuclear membrane, the complex unfolds and passes through a nuclear pore to reach the cytoplasm (Fig. 5.8). Nuclear pores are immense protein structures with eight-fold symmetry and a diameter of about 100 nanometers. Traffic through the nuclear pore is specific and regulated. Messenger RNA is specifically trans-

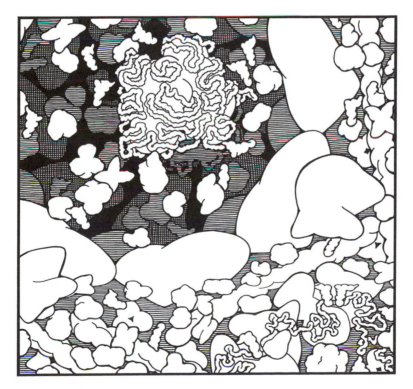

Figure 5.8   Nuclear Pore

A nuclear pore, centered at upper left, is viewed from the cytoplasmic side, looking into the nuclear interior. A complex of mRNA and protein is seen emerging through the elaborate gate of proteins. Because of the large size of the pore complex, only four of the eight symmetrical gate elements fit into the figure. (1,000,000 ×)

ported out of the nucleus, and new RNA polymerase and nucleosome proteins, which are built in the cytoplasm, but belong in the nucleus, are specifically transported into the nucleus.

TRANSPORT OF PROTEINS

With so many separate compartments, eukaryotic cells need some way of sorting and transporting new proteins to their proper location. Of course, cytoplasmic proteins are made in place by the ribosomes in the cytoplasm. Nuclear and mitochondrial enzymes, however, must be transported across membranes into their respective compartments. The large nuclear pore complex mediates the transport of protein into the nucleus, and a mitochondrial translocase protein, embedded into the mitochondrial membrane, mediates transport into the mitochondria. Other targeting tasks, for secreted proteins and for proteins of the cell surface and vacuole, are performed by the endoplasmic reticulum and Golgi apparatus. As an example, we follow the production and transport of a digestive enzyme, which will ultimately end up in the vacuole.

The journey starts at the endoplasmic reticulum, a lacelike network of tubes and sheets enclosed by a single membrane (Fig. 5.9). Special protein complexes embedded in the membrane capture ribosomes that have just started making a protein and guide the new protein, as it is being made, into the interior of the endoplasmic reticulum. Once the new protein is inside, specialized enzymes help it to fold. Other enzymes crosslink and strengthen the new proteins or destroy faulty proteins.

When the endoplasmic reticulum becomes filled with protein, parts of it are pulled off into small transport vesicles. These carry the new proteins to the Golgi apparatus, a set of membrane-bound sacs stacked like plates (Fig. 5.10). The Golgi apparatus is the processing plant of the cell. Sugars and lipids are built onto proteins that require them, and specific molecular tags are added that specify their destination. In the Golgi apparatus, proteins are modified and sorted according to their ultimate destination.

The sugar mannose-6-phosphate is the molecular tag used to mark enzymes destined for transportation to the vacuole. The sugar is connected to each protein chain, and a specific receptor protein on the inside of the Golgi apparatus binds to it. Digestive enzymes are progressively anchored to the membrane and concentrated, awaiting final transport.

When the proteins are sorted, small vesicles are removed from the edge of the Golgi body, with all of the enzymes stuck to the walls inside (Fig. 5.11). Three-armed triskelions provide the leverage to pinch off these vesicles. They

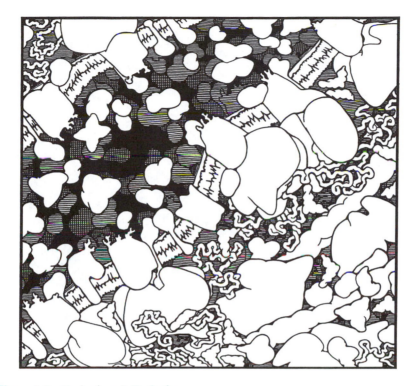

Figure 5.9   Endoplasmic Reticulum

Ribosomes bind to special transport proteins in the endoplasmic reticulum membrane, shown here in cross section. The newly built proteins are guided inside the endoplasmic reticulum (the inside of the endoplasmic reticulum looks dark in this figure because of the lower concentration of proteins than in the cytoplasm). The proteins inside have begun their journey to the vacuole or cell surface. (1,000,000 ×)

combine to form a geodesic structure on the cytoplasmic side of the membrane, pulling off spherical transport vesicles of a very uniform size. After the vesicle separates from the Golgi apparatus, the triskelions fall off and go to work on the next vesicle. Guided by microtubule rails, the free vesicle is moved to its target destination, the vacuole.

When the vesicle arrives at the vacuole, the membranes of the vesicle and the vacuole fuse, like two drops of oil in water, and the cargo of the vesicle is unloaded into the vacuole (Fig. 5.12). The acidity of the vacuole triggers the release of the digestive enzymes from their receptors; they float free and begin to work on the food and debris inside.

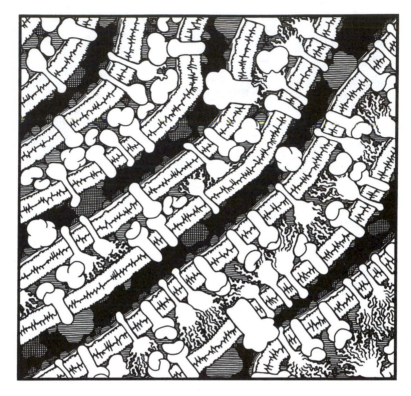

Figure 5.10   Golgi Apparatus

The stacked disks of the Golgi apparatus are seen here in cross section. As proteins pass through the stack, from upper left to lower right, they are successively modified and sorted. Long polysaccharides are added to membrane proteins and small digestive enzymes are gathered by membrane-bound sorting proteins. (1,000,000 ×)

Figure 5.11   Coated Vesicle

Reaching the last compartment of the Golgi apparatus, seen in cross section at upper left, proteins are transported to their ultimate destination in small vesicles. Three-armed triskelions (one is seen at lower left) associate into a geodesic structure on the surface of the Golgi apparatus membrane and pull off vesicles. In this figure, the process is just over half finished. The triskelions have pulled out a bubble of membrane, pulling toward lower right, and are in the process of pinching off the back face. (1,000,000 ×)

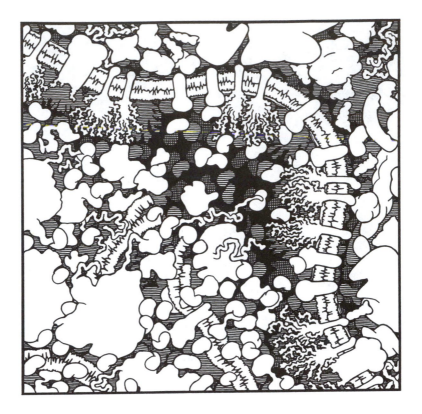

Figure 5.12   Vacuole

Reaching the vacuole, seen here in cross section, the vesicle dumps its cargo of digestive enzymes and membrane proteins. The interior of the vacuole, at lower left, is a waste heap of proteins and nucleic acids in the process of destruction. In this illustration, the vesicle has almost completely fused with the vacuole; only a dimple is left at upper right. (1,000,000 ×)

The vacuole, similar to the lysosomes of animals cells, is the trash dump of the cell. Old mitochondria and used ribosomes are pushed into it to be broken up and digested. The digestive enzymes in the vacuole are designed to perform at highly acid pH, so proteins in the membrane constantly pump hydrogen ions inside, acidifying the interior. But the acidity is deleterious to the membrane proteins of the vacuole. They are generally not as tough as the small, crosslinked digestive enzymes, so the membrane proteins are protected by a gluey coat of sugar covering their portions that project into the vacuole.

Chapter 6

Man: The Advantages of Specialization

Being multicellular has definite advantages. The earliest colonies of cells reaped the immediate profit: multicellular organisms can be built in larger and larger sizes. Large organisms tend to eat smaller organisms, not the reverse. Over the course of evolution, the meek have generally not inherited the earth, giving multicellular organisms a competitive edge.

Being multicellular has a second, more profound advantage: the freedom to specialize. Through time, individual cells in a multicellular organism may gradually focus on one aspect of living, relying on the other cells for the other aspects. Cells on the surface of an evolving organism may specialize in protection and locomotion, expending their resources on waxy coats or systems of flagella. Interior cells may focus on energy production, improving the mechanisms of absorption and digestion. A few remarkable cells may focus on reproduction, evolving into a drought-resistant seed that grows into an entirely new organism. And once a basic core of cells is in place, dedicated to using food and reproducing, the remaining cells may focus on novel ways to exploit the environment. They can specialize in making hair to protect the organism from the cold, or in sensing the level of light to find food, or in creating beautiful pigments to attract mates or warn enemies.

The form of our bodies and the levels of thought and sensation we experience developed through this freedom to specialize (Fig. 6.1). We can see the colors of a spring garden because cells in our eyes specialize in focusing and sensing light. We can take a bracing walk in the rain because large tracts of cells in our muscles specialize in forcibly changing their length. We can enjoy the sensual touch of a loved one because cells in our skin specialize in detecting small changes in their own shape. We are challenged by a riddle because cells in our brains specialize in sending, receiving, and processing electrical signals. Our bodies are composed of about two hundred different types of cells. Each type is specialized for a different task.

In spite of their diversity, however, each of our ten trillion individual cells

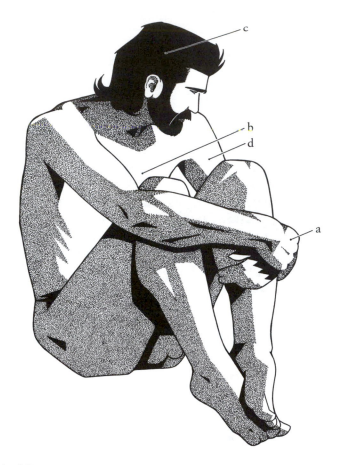

Figure 6.1   Man

Portions of four different types of cells are enlarged in the following figures (individual cells are shown in Fig. 1.2). a. Skin cells from the hand are used to illustrate chromatin in the nucleus, the cell surface, a spot desmosome, and a gap junction. b. A drop of blood from the heart is used to illustrate a red blood cell and blood serum. c. Nerve cells from the brain are used to illustrate a nerve axon and synapse. d. A muscle cell from the arm is used to illustrate the arrangement of actin and myosin.

closely resembles the simpler eukaryotic yeast cell. Each is crisscrossed with a similar cytoskeleton. Each contains a nucleus filled with DNA. Each contains a set of mitochondria, a network of endoplasmic reticulum, several Golgi apparatuses, and an array of digestive lysosomes. Our cells make their specialized proteins on top of this necessary overhead: nerve cells make the proteins for signal transmission; red blood cells make so much hemoglobin that there is not room for much else; muscle cells make actin and myosin, which together form an engine of contraction.

Our cells have an informational problem not faced by yeast. Since all of the cells in our body were created from a single fertilized egg cell, the single copy of DNA from each of our parents must hold all of the information needed by every type of cell. Thus, nerve cells have information for building hemoglobin, and blood cells could conceivably make neurotransmitters.

Clearly, this must not occur in our bodies. Each cell must be able to control which proteins it makes, focusing on its role in the blood, skin, or elsewhere. In the first nine months of life, our proliferating cells progressively commit to different fates. When a cell commits to a function, the need for a host of proteins disappears: developing nerve cells have no need for hemoglobin and blood cells no need for myosin. The DNA encoding these unneeded proteins is put into storage by the interaction of the small protein histone 1 with nucleosomes (Fig. 6.2). Histone 1 links neighboring nucleosomes together into long helical segments of chromatin, which may be densely packed in the dusty corners of the nucleus.

Our cells must also deal with structural problems that far exceed those faced by yeast cells. To build an organism as large as ourselves, elaborate molecular methods of bracing, buttressing, and connecting cells are needed. Our bodies are strengthened and shaped at many levels: from the internal bracing of cells, to strong methods of welding cells together, to tracts of resilient sugars and proteins that enclose and organize groups of cells into tissues and organs, to the flexible membranes and cartilage that attach organs to our skeleton of mineralized bone.

Each of our cells is strengthened by an internal cytoskeleton similar to that in the yeast cell. Actin filaments, intermediate filaments, and microtubules combine to form a matrix that permeates the cell, holding it in the proper shape. Our cells generally do not make a tough outer wall, like yeast, and are thus more flexible. White blood cells are a good example: like the familiar *Amoeba proteus*, they crawl through our circulatory system in search of foreign cells, which they engulf and digest. This motion would be impossible with a rigid cell wall holding the cell in one shape. Motion is made possible by an

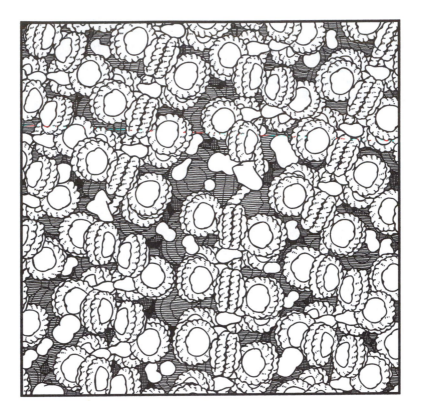

Figure 6.2   Chromatin

The small histone 1 protein arranges nucleosomes into spirals of chromatin. The DNA
in these densely packed fibers is in storage, awaiting future need. (1,000,000 ×)

internal system of bracing, which may be built and modified at the need of the
cell (Fig. 6.3). Immediately inside the cell membrane is a region of densely
packed and interconnected actin filaments that form a tough internal skeleton.
When a white blood cell wishes to move, actin filaments are broken down and
rebuilt in the proper orientation, pushing the cell in a new direction and allowing the posterior regions to retract.

Most of our cells, however, do not need to crawl around inside our bodies.
They remain fixed in tissues and organs, working with their billions of neigh-
bors to perform a particular function. A variety of junctional molecules have
evolved to hold these cells together and allow them to communicate. Spot
desmosomes glue neighboring cells together, braced by a network of intermediate filaments in each cell (Fig. 6.4). Tight junctions form a waterproof seal

between two cells, especially important in the cells lining our digestive system. Adhesion belts connect bundles of actin filaments to brace the cell wall. Together, these junctional complexes link the internal skeletons of neighboring cells, strengthening entire tissues.

Neighboring cells also communicate with one another through the tiny pores in gap junctions (Fig. 6.5). Connexons, hexagonal proteins that span the cell membrane, connect the cytoplasm of neighboring cells and allow small mole-

Figure 6.3   Cell Surface

A typical human cell membrane, seen in cross section at top, is filled with protein pumps, receptor proteins, and structural proteins. Some of the proteins are extensively decorated with polysaccharides on the side facing out of the cell, at top. A polyhedral skeleton of spectrin, the bone-shaped proteins running horizontally below the membrane, butresses the cell on the cytoplasmic side. Under the spectrin mesh, actin filaments are densely packed and rigidly interconnected by small proteins. (1,000,000 ×)

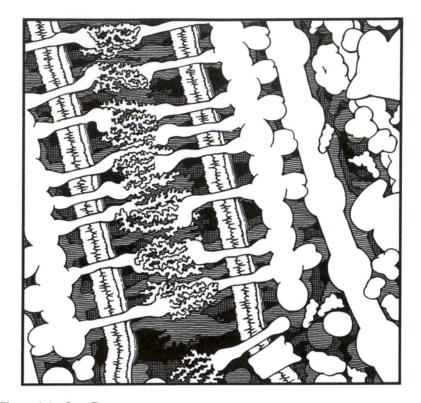

Figure 6.4   Spot Desmosome

Neighboring cells are glued together by spot desmosomes, seen here in cross section. Proteins extend across the opposed membranes, binding together in the space between. Inside each cell, a dense plaque of proteins links the transmembrane proteins to cytoplasmic intermediate filaments. The overall structure effectively links the cytoskeletons of the two cells. (1,000,000 ×)

cules to pass between. Connexons are able to open and close, although not as efficiently as some of the other cellular pumps. When the calcium level of a cell rises—often a signal that the cell is sick or compromised—connexons close and quarantine the unhealthy neighbor.

BLOOD

Blood is our waterway of transport and communication, and a major line of defense against mechanical damage and infection. When thinking of blood,

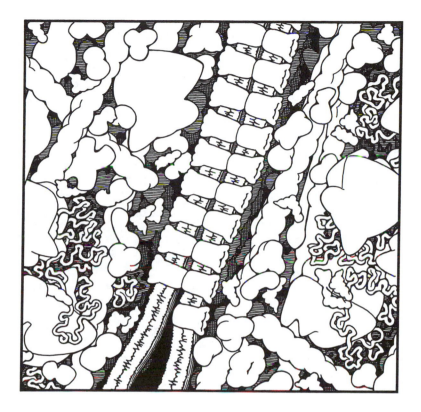

Figure 6.5   Gap Junction

Neighboring cells communicate and share resources through gap junctions, seen here in cross section. Each of the connexons has a small pore running down its center, allowing the passage of small molecules like sugar and ATP between cells. (1,000,000 ×)

our first image is of color: an embarrassed flush of color or the brightly visible accompaniment to the pain of a wound. Our five to six liters of blood is not, however, a homogeneous red liquid: more than half its volume is a clear solution of concentrated proteins. The color of blood derives from cells circulating in this clear serum, cells colored deep red by their cargo of hemoglobin.

Red blood cells are unselfishly dedicated to their work of carrying oxygen from lungs to tissues (Fig. 6.6). In fact, they can do nothing else. They are created from cells in the bone marrow. As they develop, they gradually shift all of their resources to the building of hemoglobin, allowing their other functions to atrophy. The cell membrane loses its capabilities of communication

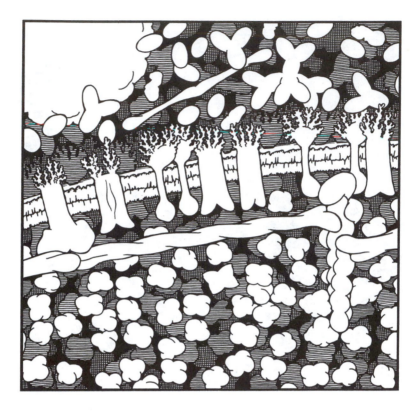

Figure 6.6   Red Blood Cell

The interior of a red blood cell, shown at the bottom, is filled almost exclusively with hemoglobin. The surface of the red blood cell, in cross section running horizontally across the center, is a simplified version of a normal cell surface. Spectrin and short lengths of actin are all that is left of the rigid cell cortex normally strengthening the inside of the cell membrane (compare this with Figure 6.3). Outside the cell, at top, is blood serum, described in the next illustration. (1,000,000 ×)

and selective transport, leaving only a simple scaffolding to help the cell hold its shape. Finally, the cell makes the ultimate sacrifice. It concentrates all of its molecular machinery—mitochondria, nucleus, ribosomes—into one portion and ejects it from its body. The mature red blood cell, now a directionless automaton, is then placed into the bloodstream, where it circulates carrying oxygen for about four months.

Blood serum is filled with a diverse collection of transport proteins designed to pick up a molecule one place and deliver it elsewhere (Fig. 6.7). The most common is serum albumin, a multifunctional blood protein that transports a number of small molecules, such as the fatty acid components of lipids. The sheer number of albumin molecules in serum is also important: if blood serum were pure water, blood cells and the cells lining the circulatory system would osmotically absorb water and burst under the pressure.

The hydrophobicity of lipids poses a special problem to the blood. The blood is the major route by which lipids are delivered between cells. But if they were merely dumped into one part of the circulatory system and picked

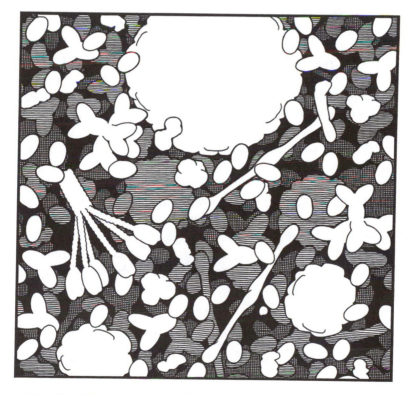

Figure 6.7   Blood Serum

Serum is a complex mixture of different proteins. At top center is one of the largest, a lipoprotein carrying lipids between cells. The most plentiful is serum albumin, drawn as small ovals. The Y-shaped proteins are antibodies, and the six-armed protein is the first protein in the complement cascade, part of our immune system. The long, rod-shaped proteins are fibrinogen, which forms the structure of blood clots. (1,000,000 ×)

up elsewhere, the way glucose is, they would aggregate into greasy clots and block the normal flow of blood. (This is approximately what happens in atherosclerosis.) Instead, lipids must be chaperoned. Lipids are transported in lipoproteins: huge molecular tankers composed of a shell of protein surrounding a globule of fat. Lipoproteins are absorbed whole into cells that need lipids and broken apart inside.

Hormones carry messages through our bodies by way of the blood. Hormones are generally small molecules, created in specific glands and received by protein receptors on the surfaces of cells. Their messages are heard system-wide, signaling cells throughout our body to respond. Epinephrine (adrenaline), created in the adrenal glands perched atop each kidney, carries a message to focus on energy production. It is often released when we are in immediate danger: the "rush" we feel is our cells mobilizing their energy resources to ready us for action. Insulin and glucagon are small protein hormones that signal the level of sugar in our blood (see Color Plate 11). Modified forms of cholesterol carry our sex-specific signals: estrogens from women's ovaries and testosterone from men's testes signal some of the bodily changes that occur in puberty. Growth hormone, a short protein made in the pituitary gland, helps regulate the growth of children. The use of hormones as messengers, however, is limited to general messages, because a different hormone is needed for each new message, as well as an entirely new set of molecular machinery to build, send, and receive it.

Blood serum also helps defend our bodies against physical damage. Damage is controlled by the formation of clots, which dam leaks until the surrounding tissue can be rebuilt. Long fibrinogen molecules are poised ready to form a fibrous clot, activated by a cascade of protein signals. Cascades are used in many cellular functions to amplify a response. For blood clotting, the cascade starts with a protein that senses damage. It in turn activates a second protein, which activates a third, which activates a fourth, which activates a fifth, which ultimately activates fibrinogen to form a clot. The advantage is that a single protein molecule at each step can activate many individual protein molecules at the next step. A single protein can start a cascade that will ultimately activate enough fibrinogen to form an entire watertight clot.

Much of the machinery of our immune system is also located in the blood serum, providing protection from invasion by foreign organisms. Antibodies in serum locate and tag the unfamiliar molecules attached to foreign organisms, targeting them for destruction (see Color Plate 14). Another cascade of serum proteins, the complement system, recognizes and destroys foreign cells. The cascade starts with C1, a six-armed protein with binding sites at each tip. When several of these arms bind simultaneously to antibodies stuck on a bacterial surface, the cascade starts, leading to the formation of membrane attack complexes that pierce holes through the invader's cell wall (illustrated as the Frontispiece).

NERVES

Hormones are a crude method of communication, allowing only the simplest of messages: "I'm hungry" or "I'm scared." Early in the evolution of multicellular animals, a more flexible and powerful solution was developed: a complex network of communicating nerve cells. The organization of nerve cells may be programmed to perform incredibly diverse functions, from simple, hard-wired reflex responses to the subtlest of thought processes. It is all a matter of arrangement and interaction.

Our nerves communicate both electrically and chemically. Electrical signals carry messages rapidly over long distances: a signal travels down the narrow axon of a nerve cell at a speed between 1 and 100 meters per second. At its destination, the message is transmitted to the next cell via a chemical signal that diffuses across the narrow synaptic cleft between successive cells.

Electrical signals are carried down the length of axons by a long chain of molecular relays (Fig. 6.8). An axon is primed for transmission by a series of chemical pumps that pump sodium ions to the outside, charging the membrane just as you would charge a battery. A series of voltage-gated channel proteins then uses the charged membrane to propagate a signal down the axon. This protein can sense changes in the voltage across the membrane: if the voltage falls, the protein channel opens up and allows sodium to rush back into the cell. A signal is initiated at the head of the nerve cell by opening a few of these protein channels, causing the local voltage difference to drop as sodium rushes into the cell. Neighboring proteins sense this drop in voltage and open as well, allowing even more sodium to rush in. The signal then propagates in a wave down the axon, as each protein short-circuits its portion of the charged membrane and the next protein responds to the change. After the wave has passed, the membrane is left completely uncharged. The channels then spontaneously close and the molecular pumps work to recharge the membrane, preparing it for the next signal.

In many of our nerve cells, particularly those which carry signals to distant parts of our bodies, the axon is insulated. Special Schwann cells wrap a multi-layered coat of insulating lipid around the conductive membrane, shielding the signal from crosstalk from neighboring axons, just as wires are insulated in telephone cables.

When an electrical wave reaches the end of an axon, it is transmitted to the

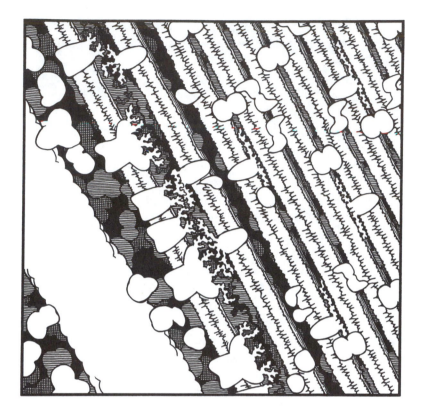

Figure 6.8   Nerve Axon

The excitable membrane of a nerve axon is shown as the leftmost membrane in this cross section. Two major proteins are embedded in the membrane. The smaller is a sodium pump, which charges the membrane. The larger is a voltage-sensing channel, which propagates the electrical signal down the axon, from top left to bottom center. Inside the axon (to the left of the excitable membrane) is the cytoplasm, including one of the large microtubules that hold the axon in its elongated shape. Insulating the excitable membrane are many concentric lipid bilayers (to the right of the excitable membrane), which are wrapped around the axon by a second cell. (1,000,000 ×)

next nerve cell via a chemical signal (Fig. 6.9). The electrical wave causes the axon terminal to dump the contents of a few hundred small vesicles into the synaptic cleft. Each vesicle contains several thousand neurotransmitter molecules, each composed of ten or twenty atoms. The neurotransmitter molecules cross the cleft and bind to specific receptor proteins on the recipient cell, causing them to open. Ions then enter the recipient cell and a signal begins once again.

One might envision a simpler solution than the chemical message of synapses. If two cells were in very close proximity, an electrical message could be transferred directly, saving the expense of making neurotransmitters. In fact, this solution is used in some cases. The chemical message of synapses, how-

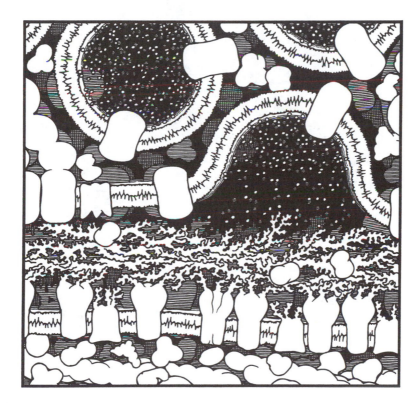

Figure 6.9   Nerve Synapse

Chemical signals are transmitted from one nerve to the next across a synapse, seen here in cross section. The end of an axon is at the top, with two vesicles full of neurotransmitters inside the axon terminal and one vesicle in the process of dumping neurotransmitters into the cleft. The synaptic cleft separating the two cells runs horizontally below center and the receiving cell is at the very bottom. Neurotransmitters cross the cleft and bind to receptor proteins embedded in the receiving cell membrane. $(1,000,000 \times)$

Figure 6.10   Muscle Contraction

The force of muscle contraction is created by the interaction of two types of protein filament: thin filaments of actin, two of which run horizontally at the center, and thick filaments of myosin, one of which runs horizontally at top and one at bottom. The many heads extending from each thick filament bind to a neighboring thin filament and, using ATP, generate the microscopic power strokes that combine to contract our muscles. (1,000,000 ×)

ever, has a definite advantage over direct electrical communication. It allows sensitive control of the response of the recipient cell. Nerve cells generally have thousands of axon terminals ending on their surface, so they receive many messages. With chemical signals, these messages can be distinguished from one another. Some of the synapses use neurotransmitters that deliver a positive, excitory message, eliciting an electrical signal in the recipient cell. Others use different neurotransmitters that deliver an inhibitory signal, suppressing an electrical response. A nerve cell may receive thousands of these chemical signals, both excitory and inhibitory, and their relative numbers determine whether an electrical signal is sent down its axon. Thus, nerve cells are processors of information, not just passive relayers of a single message.

Of course, electrical signals shooting throughout the body are not much use unless they have some physical effect. Nerves make their signals felt primarily through the action of muscles: nerves tell muscles when to contract (Fig. 6.10). Arrays of protein filaments in muscle cells create the force of muscle contraction. Thin filaments, composed mainly of actin, are surrounded by thick filaments, composed of myosin. Using ATP, the flexible heads of myosin reach out to the thin filaments, bind, and change to a shorter shape, dragging the thin filaments along. The combined action of millions of these small motions pulls the filaments relative to one another, shortening the entire cell. The combined effort of billions of these cells creates the force to move the substantial weight of our bodies.

Chapter 7

Plants: Gathering Energy from the Sun

When a lucky bacterium developed the first successful photosystem about a half billion years after the first cells arose, the organisms of the world were split into two distinct classes: providers and consumers. Until that time, all were consumers, fueling their metabolic processes with the geochemically formed organic and inorganic compounds then commonly found in the ocean. (These natural foods, provided by the earth, were used up in the early history of life. Today only a handful of organisms, such as the bacteria that utilize the hydrogen sulfide in hot springs and deep-sea hydrothermal vents, rely on them as a source of energy.) The perfection of a photosystem—a protein complex containing a series of light-absorbing chlorophyll molecules—allowed these new bacteria to create their own source of chemical energy from the inexhaustible source of the sun. The sharp division between provider and consumer has lasted to this day. Plants and photosynthetic bacteria create food out of thin air (and light). All other organisms eat plants, directly or a few steps removed.

Despite their fixed nature, and despite their role as the central providers on the earth, plants are not by any means at the mercy of their motile predators. Some protect themselves directly. Cacti clothe themselves in effective physical barriers, foxglove makes a powerful cardiac poison, and stinging nettles inject acid into attackers. Others flagrantly take advantage of animals: cherries, raspberries, and apples are nothing more than clever ways of making animals distribute seeds; flowers are made simply to be beautiful, to butterflies and bees (Fig. 7.1). Plants and animals have evolved in concert, each pushing the other to new heights of novelty in structure and function.

The photosynthetic lifestyle of plants allows them to be immobile, rooted to one place in contrast to our frenetic lifestyle. The bright green color of plants

Figure 7.1   Mountain Globemallow

A cell from the leaf of mountain globemallow (*Iliamna rivularis*) is used for the following three illustrations.

also visibly sets them apart from us: they are constantly focused on photosynthesis, never letting a photon of red light get away. The similarities between plants and ourselves, however, far outweigh these differences. Our common ancestry is apparent in every cell. The basic mechanisms of protein synthesis, cell division, and energy utilization are all but indistinguishable in plant cells and our cells.

CELL WALL

Plants have elaborated upon the rigid, external cell wall common to many microorganisms, completely committing themselves to a life of immobility. Each plant cell lives in a strong but resilient box. Plants use their individual enclosures for support: whereas our bodies are supported by mineralized bones, plants are held up by inflating their cells. Water circulating around plant cells, drawn from roots to leaves, osmotically seeps into each cell, creating an internal pressure up to 50 times that of the surrounding atmosphere. If an animal cell were placed in this position, it would absorb water until it burst. The plant cell, however, is bounded by a tough box. The cell swells, but the wall keeps it from bursting. The resulting pressure is used to hold the plant erect. But if the plant doesn't get enough water, its cells lose pressure, and the plant wilts.

The cell wall is composed primarily of polysaccharides (Fig. 7.2). Cellulose is the most abundant structural element. In fact, it is the most abundant molecule made by the organisms of the earth. The long strings of glucose molecules in cellulose form rigid rods aligned along the cell surface. The cellulose fibers are embedded in a matrix of other polysaccharides to form the cell wall, much as the reinforcing rods in bridges are glued together with concrete. These include pectins, branched polysaccharides commonly used to thicken jellies and jams.

The resulting wall is impenetrable to large molecules, such as proteins and nucleic acids, but allows the percolation of small molecules, such as water and salts. For this reason, plant hormones tend to be very small molecules, so that they are able to move from cell to cell at a useful rate. Gibberellins, auxins, and other growth-regulating plant hormones are all composed of twenty or thirty atoms. The hormone that controls fruit ripening is even smaller: the gas ethylene, composed of only six atoms.

PHOTOSYNTHESIS

Photosynthesis is performed in small, membrane-enclosed chloroplasts, compartments about the size of mitochondria. Like mitochondria, chloroplasts are thought to be visitors to the cell. Sometime during the early evolution of eukaryotic cells, a photosynthetic bacterium invaded a cell, or was eaten, and came to make the inside of this cell its home. Modern chloroplasts contain their own set of DNA and the molecular machinery for its transcription, translation, and duplication. Chloroplasts make about 120 of their own proteins,

Figure 7.2   Cell Wall

Strands of cellulose, running horizontally at top, are glued together by branched poly-saccharides such as pectin. The cell membrane, running horizontally below center, contains a variety of functional and structural proteins. Inside the cell, at bottom, a microtubule helps to brace the cell membrane from the inside. (1,000,000 ×)

the rest being provided by the host cell. Like mitochondria, chloroplasts reproduce by breaking in two and passing into daughter cells along with the cytoplasm.

Chloroplasts contain a chemical battery that is charged by light, and mechanisms to convert its energy into organic foodstuffs. The battery is a stack of thylakoid disks: flat, membrane-enclosed sacs that hold the proteins of the light reactions (Fig. 7.3). (Photosynthesis is described in Chapter 3.) Photosystems are the most plentiful protein complexes embedded in these membranes, absorbing light to fuel the transfer of electrons from water to NADPH. (All of the oxygen we breathe is a waste product of this process!) The favorable

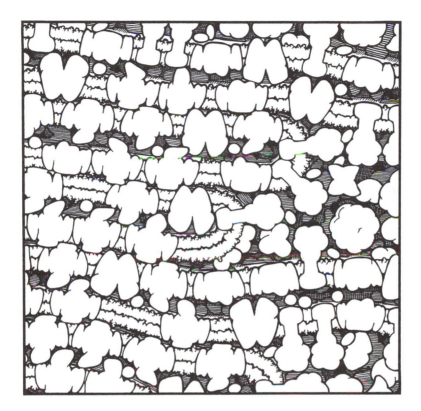

Figure 7.3   Thylakoid Disks

The light reactions of photosynthesis occur in the stacked membranes of the thylakoid disks, shown here in cross section. Photosystems are embedded side by side in these membranes, capturing the energy of light. The membranes also contain the other protein complexes in the chain of photosynthetic electron transfer. (1,000,000 ×)

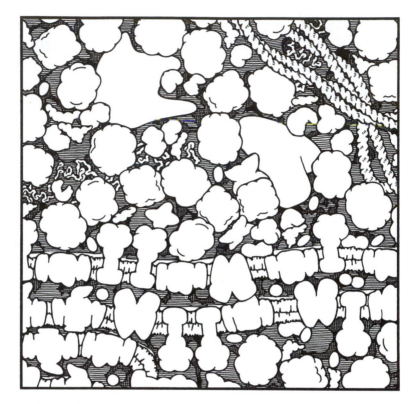

Figure 7.4   Chloroplast Stroma

The dark reactions occur in the liquid surrounding the thylakoid disks, the stroma, shown in the upper two-thirds of this illustration. The large, plentiful protein is ribulose bisphosphate carboxylase, which catalyzes the key reaction that captures carbon dioxide. Notice the presence of DNA, RNA, and ribosomes. (1,000,000 ×)

energy of the electron transfer is used to charge the membrane, concentrating hydrogen ions in the narrow interstices of the thylakoid membranes. The charge is then used to create ATP, by a protein complex similar to that in mitochondria.

The dark reactions occur in the stroma of the chloroplast, the liquid surrounding the thylakoid disks (the original cytoplasm of the ancestral photosynthetic bacterium) (Fig. 7.4). The most prevalent enzyme is ribulose bisphosphate carboxylase, which uses ATP to fix carbon, converting useless carbon dioxide into biologically useful sugar.

III: Cells in Health and Disease

A living cell is a delicate, interconnected assemblage of molecular machines and processes. There is not a great deal of room for error. If a necessary vitamin is not available, the cell dies. If a virus injects a malignant piece of nucleic acid, the cell dies. If a poison destroys one key enzyme, the cell dies.

Chapter 8

Vitamins

Vitamins are vital links in our metabolism: molecules that our body absolutely needs, but that our cells cannot make themselves. Somewhere in the evolution of our distant ancestors, we lost the ability to make these molecules, presumably because they were always available in the diet. Today we have no choice. We must obtain these important molecules in our food.

A rich body of folklore prescribes certain foods for specific ailments: carrots or liver for night blindness, cod liver oil for rickets, and rose hip tea or limes for scurvy. Often, folk cures can be traced to the vitamin content of certain foods. Scientific analysis of these healing foods has revealed that specific molecules in each are responsible for the cure. The first characterized, thiamine, gave vitamins their name: amines that are vital for life.

VITAMIN A (RETINAL)

Retinal, commonly known as vitamin A, is the molecule that senses light in our eyes (Fig. 8.1). Foods rich in retinal, most notably liver, have been prescribed since antiquity for night blindness. Eating the liver of a fish or bird improved the vision of seamen virtually overnight. Retinal is composed of a string of carbon atoms that readily absorbs the energy from a photon of light. When it absorbs light energy, retinal changes shape from bent to straight, a difference large enough to be easily sensed by opsin, a membrane-bound protein. Opsin binds a molecule of retinal at its center; when retinal changes shape, opsin in turn triggers a nerve signal telling the brain that the cell has just seen a photon of light.

Vitamin A is very hydrophobic and dissolves readily in fat. This has important consequences in our bodies. Excess vitamin A is not excreted, as are vitamins that are soluble in water. It concentrates in the fatty parts of the liver, reaching toxic levels if too much is allowed to build up. In a normal diet there is no problem, but when large doses of vitamin A are used to treat acne or prevent colds, the chance of poisoning is significant.

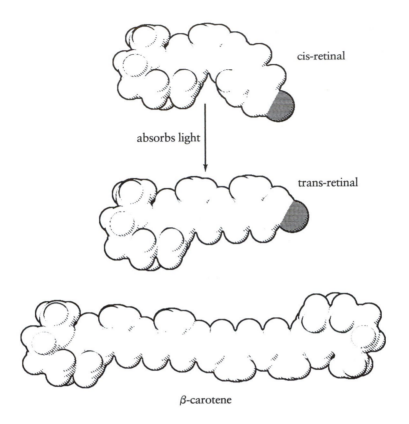

cis-retinal

absorbs light

trans-retinal

β-carotene

Figure 8.1   Vitamin A

Vitamin A is almost exactly the same as one half of the yellow pigment in carrots, β-carotene. Light causes vitamin A to change shape from the bent *cis* form to the straight *trans* form. Our eyes sense light by "seeing" this change in shape. (30,000,000 ×)

The European folk cure for night blindness is the carrot. Carrots and other yellow and dark green vegetables contain molecules just as good as retinal: bright yellow and orange carotenes, twice the size of retinal. Enzymes in our cells cleave carotene molecules into usable molecules of retinal.

B VITAMINS

Several vitamins are used in our bodies to build special carrier molecules that shuttle hydrogen, nitrogen, or carbon atoms between various enzymes. These vitamins fall collectively into the B complex (Fig. 8.2). We generally

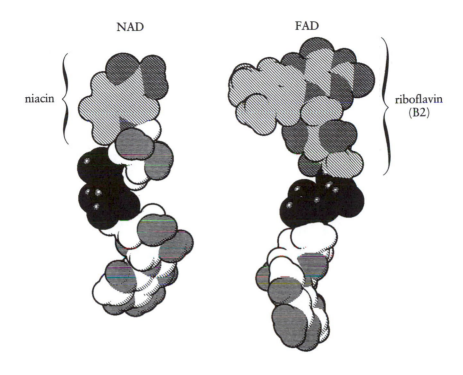

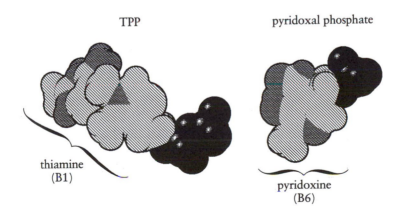

Figure 8.2   B Vitamins

The B vitamins are used to make carrier molecules in our cells. The portion of each carrier molecule derived from the vitamin is highlighted with diagonal lines. (30,000,000 ×)

obtain sufficient supplies of the B vitamins from milk, meat, and eggs, and from fortified cereals.

Thiamine (vitamin B_1) was the first discovered, as the cure for the fatal illness beriberi. This disease struck people in eastern Asia in the early part of the century, soon after they began to eat rice with the unsightly brown husks removed (polished rice), in the best European tradition. This disease affects the entire body, starting in few days with mental confusion and lack of coordination, and leading to advanced neuromuscular problems and buildup of liquid in the tissues. Such overwhelming symptoms highlight the central role played by thiamine.

After two phosphates are added to one end to form thiamine pyrophosphate (TPP), thiamine is used by several key enzymes. Two enzymes in the central path of energy production rely on thiamine: the step after glycolysis (performed by the large enzyme complex pyruvate dehydrogenase), and a similar reaction in the citric acid cycle. Thiamine has a particularly reactive carbon atom that is used by these enzymes to coax a carbon dioxide molecule away from the molecules being broken down.

Riboflavin (vitamin B_2) and niacin are needed to make the two carrier molecules that shuttle hydrogen atoms around the cell: riboflavin for FAD (flavin adenine dinucleotide) and niacin for NAD (nicotinamide adenine dinucleotide). Niacin is not technically a vitamin because our cells can make it from the amino acid tryptophan. But its synthesis is inefficient. The plentiful sources of niacin in our diet are adequate to fill our metabolic needs, so our cells usually don't bother to make it themselves.

Pyridoxine (vitamin B_6) plays a central role in the metabolism of amino acids. In the cell, it is converted to pyridoxal phosphate by adding a phosphate group. Pyridoxal phosphate, like thiamine, has a particularly reactive carbon atom, which is used in many enzymatic reactions. It is needed by enzymes that transfer nitrogen atoms in reactions where amino acids are synthesized or broken down. Vitamin B_6 is also needed by several enzymes that make the neurotransmitters norepinephrine and serotonin. The symptoms of vitamin B_6 deficiency are usually caused by the lack of these neurotransmitters, starting with nervousness and depression and leading to severe nervous ailments.

VITAMIN C (ASCORBIC ACID)

Ascorbic acid, commonly known as vitamin C, has a simple structure that belies its status as perhaps the most controversial of vitamins (Fig. 8.3). Its hotly debated efficacy against colds has been well tested. Large doses appear to reduce the severity of colds, but not their frequency. Vitamin C also appears

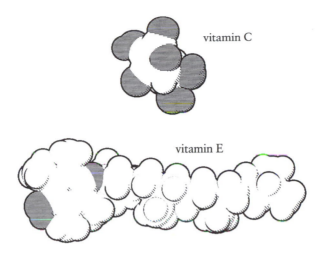

vitamin C

vitamin E

Figure 8.3   Vitamins C and E

Vitamins C and E act as antioxidants in our cells. Vitamin E has a long hydrophobic tail, so it performs its function dissolved in lipid membranes. The more water-soluble vitamin C is found in body fluids, protecting water-soluble molecules. (30,000,000 ×)

to reduce the risk of cancer, as do other antioxidants such as the carotenes mentioned above with vitamin A. Antioxidants help protect cells from the ravages of oxygen, which can destroy lipids and proteins.

The major function of vitamin C in our bodies has, however, remained somewhat elusive. Its most visible role is in the formation of collagen, the structural molecule in cartilage. This explains the unpleasant symptoms of scurvy: loss of teeth, slow healing, and hemorrhaging. Many folk cures exist to avoid the ravages of scurvy when citrus fruits are not available: Scandinavians drink rose hip tea and Eskimos eat raw fish.

However, only a small amount is needed for this function, making it difficult to explain the high concentrations of vitamin C found in both animal and plant tissues. Perhaps its action as an antioxidant, protecting our molecules from oxygen, is its major cellular role.

VITAMIN D

Vitamin D is used as a hormone in our bodies to regulate the uptake and release of calcium in our bones (Fig. 8.4). Vitamin D can be made in our bodies, as long as we are exposed to the sun. In our skin, ultraviolet light makes vitamin D by breaking one bond in a form of cholesterol.

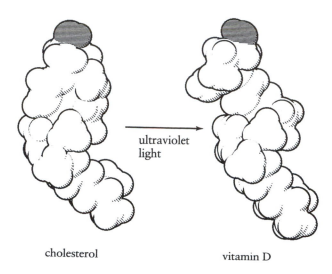

cholesterol   vitamin D

Figure 8.4   Vitamin D

In our skin, a bond in cholesterol is broken by ultraviolet light, forming vitamin D. In the liver and kidneys, several oxygen atoms are then added, forming a hormone used to regulate calcium levels. (30,000,000 ×)

Problems occur when people cannot get into the sun, particularly people who live at high, cloudy latitudes. They cannot make enough of the vitamin themselves and must get it in supplements, such as the infamous doses of cod liver oil forced upon English children. Children born in the winter are especially susceptible. They go without this necessary vitamin for the first few months of their life, developing the poorly formed bones characteristic of rickets, the human equivalent of a rickety chair.

VITAMIN E (TOCOPHEROL)

Vitamin E is another controversial vitamin (Fig. 8.3). Its role in our bodies is still obscure. It is a powerful antioxidant like vitamin C. Unlike vitamin C, however, it is soluble in fat because it is largely hydrophobic. Therefore, the major role of vitamin E is most likely the protection of lipids from oxygen. The oxidation of lipids is highly undesirable: when butter and grease go rancid, they are actually being oxidized by oxygen in the air. Cells do not want their cell membranes to go rancid.

The name tocopherol comes from the Greek word for childbirth, because it was found that a vitamin E deficiency in mice leads to sterility. This effect is not found in humans. In fact, the symptoms of vitamin E deficiency in humans are not precisely known.

Chapter 9

Viruses: Biological Highjackers

Imagine that you want to force a cell to make a certain protein in large quantity. What could you do to subvert the cell's normal operations to do your bidding? It might be enough to inject a short piece of DNA that encodes the protein. The molecular machinery in the nucleus would then transcribe the foreign DNA into mRNA, which would in turn be translated into your protein in the cytoplasm. This is how many genetic engineering companies currently coax bacteria and yeast cells to make large quantities of insulin and growth hormone for medical applications.

This process is not overly efficient, however, because the DNA you inject must compete with all of the other normal cellular DNA. One way to get around this problem is to inject a special polymerase along with the DNA designed to transcribe RNA only from your short piece of DNA, and not from the cell's own DNA. The cytoplasm will then become enriched in the RNA coding for your protein, and thus more of your protein will be made. Next, inject a second enzyme along with the DNA and the polymerase, an enzyme which seeks out and destroys the cell's normal polymerases. With these two enzymes, all of the cell's normal protein synthesis is stopped. Only your selected DNA is transcribed, and only your protein is made. The cell fills up with large amounts of your protein, but the cell eventually dies when its stored resources are used up.

A subtle refinement of this method is possible. Inject instead a single DNA molecule that encodes all three proteins: your protein of interest, the polymerase, and the enzyme of destruction. When this longer DNA is injected, the cell's machinery makes all three proteins as well as doing its normal work. But when the new transcription and inhibitor proteins reach high enough levels, normal cell transcription is stopped, leaving the cell to make only your proteins. True, the cell makes more polymerase than necessary, but one piece of DNA was all that was needed.

This is exactly what viruses do. Viruses inject a strand of nucleic acid into cells (DNA in some viruses, RNA in others) which directs the formation of a new polymerase and proteins that shut down other cellular processes. The proteins that viruses want to make, analogous to the insulin and growth hormone desired by genetic engineering companies, are the proteins making up their own "bodies." Viruses completely hijack cells: after making the polymerase, infected cells are forced to make millions of new viruses. When the resources of their host are spent, the new viruses burst out of the cell, ready to inject their nucleic acid into other healthy cells.

Viruses cannot, however, be called living (Fig. 9.1). The simplest are composed of a single strand of nucleic acid surrounded by a thin coat of protein, the entire package barely the size of a ribosome. Remarkably, this is all that is needed to completely hijack a cell. But without the molecular machinery of a cell, viruses are nothing. They cannot reproduce on their own. They cannot

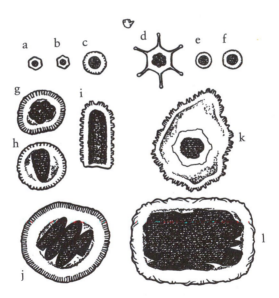

Figure 9.1   Viruses

Viruses have many shapes and sizes. The simplest are composed of a simple protein coat enclosing a strand of nucleic acid: RNA in poliovirus (a), rhinovirus (cold virus) (b), and rubella virus (c), and DNA in adenovirus (d), hepatitis virus (e), and papilloma virus (warts) (f). Other viruses are surrounded by a lipid bilayer studded with proteins, with a complex array of proteins and nucleic acids in the center. Some which contain RNA are influenza virus (g), human immunodeficiency virus (AIDS virus) (h), rabies virus (i), and mumps and measles viruses (j). Others which contain DNA are herpes virus (k) and smallpox virus (l). A bacterial ribosome is shown at top for size comparison. (100,000 ×)

make any of their own component parts—amino acids and nucleotides. They are blind assailants, inert until they bump into the side of a living cell.

POLIOVIRUS AND RHINOVIRUS

Poliovirus, the agent of poliomyelitis, and rhinovirus, the main cause of the common cold, are among the simplest viruses that attack human beings. They are composed of a single strand of RNA, about 7500 nucleotides long, encapsulated in a hollow protein coat. The coat is made of four different types of protein, which associate like panels in a geodesic dome. These viruses are so uniform that they can be coaxed to crystallize, allowing researchers to determine their exact atomic composition.

The polio vaccine ranks among the great triumphs of modern medicine, whereas the lack of a cure for the common cold is an oft-voiced complaint of irate taxpayers. The reason for this striking contrast is simple and intrinsic to the viruses themselves. There are only three strains of poliovirus, each with slightly different coat proteins. Our bodies require three different types of antibody to recognize and destroy any poliovirus they may encounter. Supported by an incredible public fundraising effort led by the March of Dimes, Jonas Salk and Albert Sabin developed the vaccine now in widespread use. (Government funding of biological and medical research is relatively new; in the past, most research required private support.) Vaccines are composed of actual virus particles, inactivated or weakened by chemical means, but still able to stimulate the body into making antibodies to prepare for a real attack. The trick is to include all three viruses in the vaccine, so the body makes antibodies against all three at the same time.

Herein lies the problem with a cold vaccine. Over one hundred strains of rhinovirus that cause colds are known. Each of the colds we suffer through is due to a strain that we have never encountered before and that we therefore have no antibodies against. The population of cold viruses is too heterogeneous to permit an effective vaccine.

A difference in the protein coats of these two viruses accounts for their distinct clinical effects. The coat of poliovirus is stable in acid, whereas that of rhinovirus is unstable. When the viruses are ingested, by eating contaminated food or breathing infected aerosols, the viruses infect different places. Poliovirus is perfectly happy in the acid environment of the stomach, so it may infect cells there and spread through the lymph to the rest of the body. Rhinovirus, however, is destroyed in the stomach and therefore has most of its effect above, in the throat and nose. The instability of rhinovirus keeps it confined to inconvenient places, while the stability of poliovirus allows it to reach more vital regions of the body. Most people infected with poliovirus experience only a bad fever, but in about 1% of infected people, the virus takes a stronger hold, causing serious damage to nerve cells and the brain. Providence has provided, fortunately, that the more effective agent of disease is also the one we can effectively control.

Poliovirus and rhinovirus are very small and can only carry a limited amount of information with them. The strand of RNA in poliovirus is long enough to code for about ten proteins. All of its functions must be carried out with only these ten or so proteins. There is simply no room for more. The first protein needed by poliovirus is a special polymerase that creates RNA using RNA rather than DNA as a template. Second, the poliovirus RNA needs to be differentiated from the normal RNA of the host cell. Poliovirus tags its RNA by attaching a small protein to one end. All newly made RNA molecules have this protein attached to one end soon after they are formed.

The third protein built by poliovirus is a protease, an enzyme that breaks protein chains, similar to our digestive enzymes. Since poliovirus RNA is one long piece, it is translated into one long protein chain that must be broken into the ten or so separate proteins. Some of these cleavages occur by themselves, but others need help. Therefore, the third protein is a specific protease that makes the necessary breaks at the proper positions. Four additional proteins are needed to build the coat around the intact virus, bringing the count to seven proteins. The exact functions of the remaining three or so proteins are not well understood. They presumably act in shutting down the cell's synthesis of its normal proteins.

These few proteins orchestrate a life cycle that ends in the death of the cell (Fig. 9.2). The cycle starts with viruses floating in the liquid around the target. The surface of the virus has a network of pockets that recognize and bind to glycoproteins on the surface of cells. In a process which is not well understood, the coat proteins are triggered by this binding to rearrange, forcing the RNA inside the cell.

Once the RNA is inside, the cell's ribosomes translate the RNA into a long protein, which breaks into several pieces. One piece, the polymerase, proceeds to create more RNA molecules, using the cell's reservoir of nucleotides. Other protein pieces destroy the normal transcription apparatus of the cell. As the number of viral RNA molecules rises and increasing amounts of coat proteins are made from them, new viruses begin to form. Each contains one of the new RNA molecules packed inside a coat of newly built proteins. The viral polymerase molecules and other enzymes are left behind. There is just not enough room in the virus for anything but RNA. Finally, the cell ruptures and the new viruses swarm to other cells to begin the cycle again.

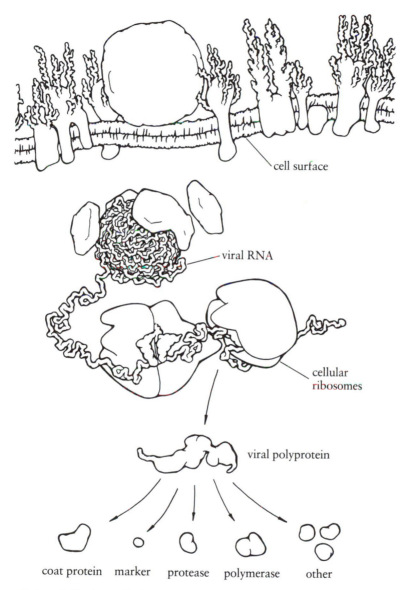

cell surface

viral RNA

cellular
ribosomes

viral polyprotein

coat protein   marker   protease   polymerase   other

Figure 9.2a   Poliovirus Life Cycle

Viruses recognize and bind to glycoproteins on a cell surface (top). Once the viral RNA is inside, the cell's ribosomes build a long viral protein (center), which is broken into several functional pieces (bottom).

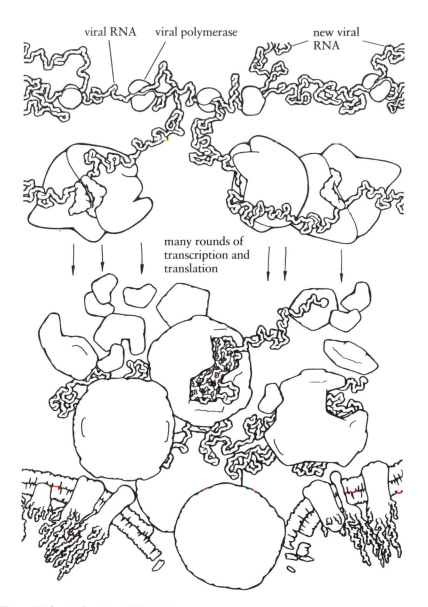

viral RNA   viral polymerase   new viral RNA

many rounds of transcription and translation

Figure 9.2b   Poliovirus Life Cycle

The new polymerase molecules then build more RNA from the original viral RNA (top). Repeated rounds of RNA synthesis and synthesis of viral protein build up large quantities of RNA and coat proteins (center), which associate into mature viruses and burst out of the cell (bottom). (1,000,000 ×)

INFLUENZA VIRUS

The virus that causes influenza is more complex than poliovirus and rhinovirus, and consequently is even more difficult to eradicate. Influenza has taken a great toll in human life. Most recently, twenty million people died in a pandemic after the First World War, more people than actually died in the war. The lasting effectiveness of influenza virus as a pathogen lies in one aspect of its structure: its genetic information is carried in eight separate strands of RNA, each coding for a few proteins. As with the cold virus, influenza virus occurs in many strains. But the segmented RNA of the influenza virus give it an extra advantage. If two different viruses infect the same cell, the separate RNA molecules may mix up. New viruses will be made with parts from each original virus, and some may end up with the best parts from each. Influenza is constantly shuffling its genes around, making it different with each bout of infection. The antibodies you make for a current infection will most likely be useless against the next one.

Influenza virus is larger and more complex in structure than poliovirus and rhinovirus (Fig. 9.3). It is coated with a lipid bilayer studded with proteins that recognize the surfaces of cells to infect. This lipid bilayer is not made by the virus: it is stolen from the infected cell. The last steps of the assembly of the virus occur at the surface of the cell. Proteins assemble on the inner face and push out buds. When the new viruses separate, they carry with them a coating of cell membrane.

HUMAN IMMUNODEFICIENCY VIRUS (AIDS VIRUS)

The AIDS virus is even more insidious than the viruses discussed above. AIDS virus is a retrovirus, characterized by a frighteningly effective difference in lifestyle. The virus carries two novel enzymes: a reverse transcriptase, which makes a strand of DNA using the viral RNA as a template (the reverse of the normal DNA to RNA transcription), and an integrase, which allows this DNA to enter the nucleus of our cells and splice itself into the cell's normal DNA. Imagine the consequences. The viral DNA is then virtually indistinguishable from our own DNA. It is duplicated along with our normal DNA when our cells divide. It is repaired just as our normal DNA is repaired. It may lie silent for years or decades after the initial infection. The virus is therefore difficult to combat. It must be recognized and destroyed before it enters a cell, for it is virtually invisible once it slips into the cellular DNA. Many of the drugs used to combat AIDS, such as azidothymidine (AZT), work by inactivating the reverse transcriptase, stopping the virus before the viral DNA is made.

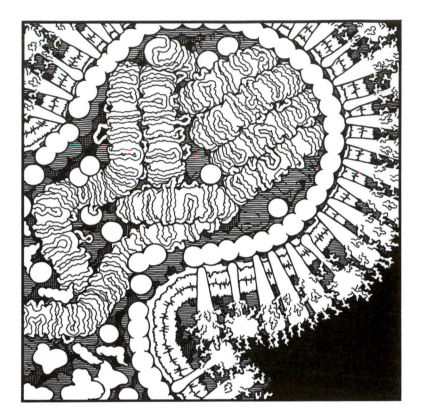

Figure 9.3   Influenza Virus

Influenza virus buds from the surface of cells, instead of bursting cells like polio-virus. In this cross section, the interior of the cell is at bottom left and a budding virus fills most of the frame. The viral nucleic acid is associated with proteins, forming large helical bundles inside the virus. The proteins lining the inside of the membrane help to force out the bud, and the proteins radiating to the outside help the virus recognize and bind to the cells they attack. (1,000,000 ×)

The AIDS syndrome occurs when the viral DNA emerges years later, often when the host cells are stressed or compromised by other disease. It separates from the normal cellular DNA, then functions much like the nucleic acid of other viruses. Many copies of the viral RNA are made, and from them, coat proteins and the two special viral enzymes are built. Finally, the virus buds from the surface of the cell like influenza virus, carrying away its RNA and a few molecules of reverse transcriptase and integrase encased in a sheath of lipid, ready to infect another cell.

Herpesvirus, the cause of genital herpes and cold sores, has a similar lifestyle. The virus preferentially enters nerve cells adjacent to the site of infection. There it incorporates itself into the cellular DNA and lies dormant. Certain chemical signals, typically those circulating in the body in times of stress or fatigue, cause the virus to emerge. It then travels down the nerve fiber to the body surface (usually a mucous membrane such as the inside of the lips) where it attacks skin cells in a typical viral fashion: entering cells, reproducing, and breaking the cells when done. The destruction of cells causes the painful sores. Later, our defenses get the breakout under control and the sore heals. But the virus still remains dormant in the nerve cells, poised for the next attack.

The AIDS virus is particularly effective and hard to combat for another reason. The AIDS virus commonly infects the cells that make antibodies, destroying one of our body's major lines of defense at its source. The long latent period and the destruction of the immune system combine to make AIDS one of the major health threats of our time. It is fortunately a very fragile virus, destroyed by the mildest of conditions outside the body. For this reason, it is spread only by the direct transfer of bodily fluids, never leaving the warm environment inside our bodies.

Chapter 10

Poisons and Drugs

Beneficial drugs and hazardous poisons are shades of one another. The old adage is entirely true: one man's passion is another man's poison. Poisons attack and destroy a key molecular machine, blocking one of the vital processes of life and killing cells. Antibiotic drugs do exactly the same thing: they block or destroy some key molecular machine. Antibiotics, however, are designed to poison only a microorganism trying to invade your body, without poisoning you at the same time. Antibiotics are specific poisons. They must selectively poison a molecular machine which is vital to the pathogenic organism, but which is not needed in our bodies.

The drugs we take to cure headache or relieve anxiety are also poisons, but much milder poisons than antibiotics or the classic historical poisons. They temporarily poison a molecular machine of lesser importance, stopping transmission of pain or modifying nerve signals. The effect of these drugs wears off in a short time, when our bodily defenses remove and destroy them.

BROAD-SPECTRUM POISONS

Poisons that block the central process of energy production will kill almost any organism they contact. Since most modern organisms use virtually identical molecular machines to transform sugar into usable energy, a poison that blocks energy production in a bacterium will be just as effective in fungi, plants, and animals. These poisons are often chemically very simple and easy to make. Hence they have played a colorful role in human history.

Cyanide is one of the most effective poisons known (Fig. 10.1). In minutes, cyanide will kill any organism that uses oxygen. It is, of course, useless as a therapeutic drug, because it attacks a molecular machine which is absolutely essential to all air-breathing organisms—bacteria, plants, and humans alike. Treating an infected patient with cyanide would kill both the pathogenic organism and the patient. Cyanide attacks the last step of energy production, when hydrogen atoms are added to oxygen to form water. The large enzyme

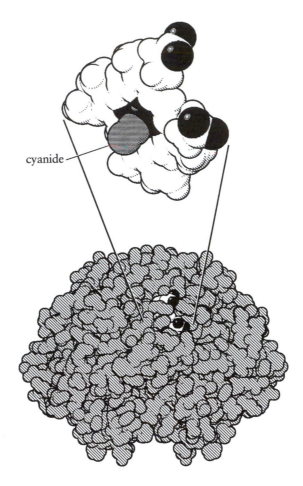

cyanide—

Figure 10.1   Cyanide and Carbon Monoxide Poisoning

Cyanide and carbon monoxide bind to iron atoms in proteins at the site where oxygen normally binds. Hemoglobin, shown at bottom, is a common target. It contains four iron atoms, each tightly held at the center of a flat heme group. Cyanide bound to the iron in one heme is shown at top. (top, 30,000,000 ×; bottom, 10,000,000 ×)

complex that performs this reaction, found in the cell membranes of bacteria and in the mitochondria of higher organisms, contains an iron atom held in a heme group, like that of hemoglobin. Cyanide binds tightly to the iron in the position normally occupied by oxygen, so oxygen is blocked from binding and entering the reaction. The cell suffocates—surrounded by oxygen, but unable to use it.

Cyanide has an infamous history. Poisonous in both its solid, salt form and as the gas hydrogen cyanide, it was discovered by the chemist Karl Wilheln Scheele, who was promptly killed by its vapors. Fast and painless, it has been used in gas chambers both in modern capital punishment and in the Holocaust. The cyanide in peach and apricot pits was used by the Greeks and Romans. The pits contain amygdalen, marketed as Laetrile for cancer treatment, which releases cyanide when exposed to alkaline conditions, as in our upper intestine. A lethal dose can be obtained from about 100 grams of ground peach pit.

Carbon monoxide is similar in shape and action to cyanide, binding tightly to iron atoms. The major site of carbon monoxide poisoning is the iron atoms of the hemoglobin. (Because hemoglobin is found primarily in vertebrates, carbon monoxide is a more specific poison than cyanide.) It binds to hemoglobin 250 times more tightly than oxygen and blocks the flow of oxygen from our lungs to the cells in our bodies. This tight binding makes carbon monoxide difficult to remove. An hour of breathing pure oxygen will reduce the level of bound carbon monoxide by only one-half.

Arsenic is another effective poison, attacking a wide range of organisms at a step in glycolysis (Fig. 10.2). Arsenic acts in the form of arsenate ion (one arsenic atom bound to four oxygen atoms). Arsenate is chemically similar to phosphate ion and acts by replacing it in cellular reactions. For example, a phosphate ion is normally taken from solution and attached to a sugar in the sixth step of glycolysis, and then used to make ATP in the seventh step. Arsenate replaces phosphate in this reaction and is attached to the sugar instead. Unfortunately, it almost immediately falls back off, because the sugar–arsenate bond is easily broken by water. Glycolysis then continues with its normal eighth step, but the cell has not made any ATP in the seventh step. Arsenate thus acts as an uncoupler: sugar is still broken down, but it is no longer coupled to the production of ATP.

Arsenic is the archetypical homicidal poison. It is commonly available in the form of rat poison. Virtually tasteless, a minuscule one-tenth of a gram of the white powder is fatal. Fortunately, it is easily detected in victims by modern forensic methods, so intentional arsenic poisoning is now uncommon. Not so in the past; it has been used in this capacity since at least Egyptian times. The famous poisonings by the Borgias and the Medici are both thought to have been arsenic. An interesting fact, often used by mystery writers, is that a tolerance for arsenic can be achieved through the administration of small doses over a period of months. The well-prepared murderer then confidently consumes the poisoned dessert along with the victim. The tolerance is presumably due to the production of extra enzyme molecules in the criminal's body, allowing it to handle the dose of poison and still leave enough enzymes functioning.

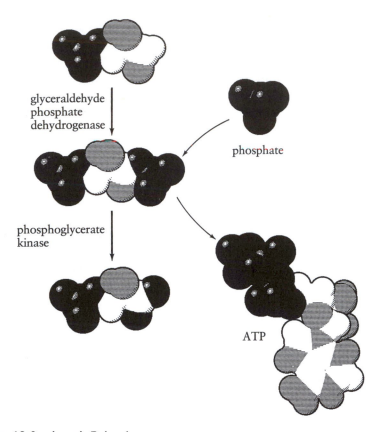

glyceraldehyde
phosphate
dehydrogenase

phosphate

phosphoglycerate
kinase

ATP

Figure 10.2   Arsenic Poisoning

Arsenate ion is chemically similar to phosphate ion and may take its place in chemical reactions. Above are the normal sixth and seventh reactions of glycolysis, which form a molecule of ATP. On the facing page are the same reactions, with arsenate replacing phosphate. Arsenate falls off in the seventh step, without forming ATP. (30,000,000 ×)

POISONS AND DRUGS OF THE NERVOUS SYSTEM

The transmission of signals in our nervous system offers many sites for the action of drugs and poisons (Fig. 10.3). Plants are especially adept at making poisons that block different steps of nerve transmission. These poisons are specific for animals. Plants have no nerves, so they can make and store the poisons in their tissues with no ill effects. Less virulent versions of these poisons are used as drugs. They transiently modify the action of certain classes of nerves, relieving tension, blocking pain, or creating recreational hallucinations.

Curare, atropine, and hemlock are plant-derived poisons, all attacking the

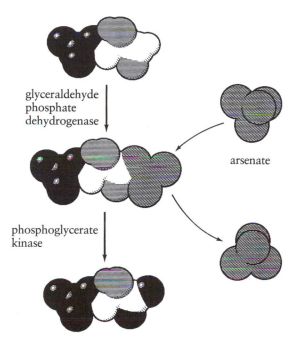

glyceraldehyde
phosphate
dehydrogenase

arsenate

phosphoglycerate
kinase

Figure 10.2

same protein. They block the binding of acetylcholine to its receptor protein
on the surface of muscle cells. Acetylcholine carries a chemical signal across
the synapse from nerve to muscle, signaling the muscle to contract. Hence,
blocking the acetylcholine receptor causes paralysis.

Native South Americans use curare in their arrows and blowguns. Curare is
made from the leaves of the bush rope (*Strychnos toxifera*), a vine that climbs
and strangles trees in the South American jungle. Early explorers brought back
a colorful tale of the concoction of curare. Apparently, several old women of
the tribe were shut in a hut with all of the necessary ingredients. The doors
were opened several days later, and if the women were not lying half-dead
from the fumes of distillation, they were punished, for the poison would not
be strong enough.

Curare is an excellent poison for hunting, because it is not absorbed through
the digestive tract. Small doses of curare will kill when injected by arrows or
darts, but the poisoned meat may be safely eaten. Curare has found modern
use as a muscle relaxant in surgery.

Atropine is a poison from belladonna (deadly nightshade, *Atropa bella-donna*), which is also found, surprisingly, in mandrake (*Mandragora officina-*

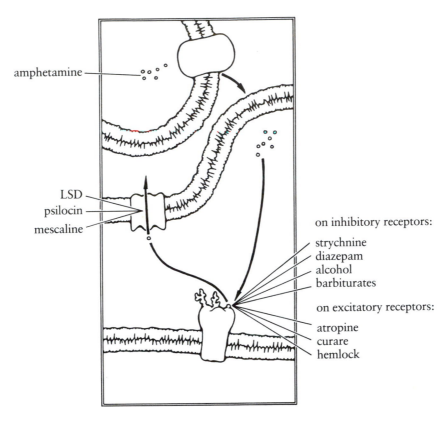

Figure 10.3   Sites of Action of Nervous System Drugs

Many drugs act at nerve synapses (see Fig. 6.9 for a more complete illustration of a synapse). Amphetamine displaces neurotransmitters out of storage vesicles. Hallucinogens such as LSD, psilocin, and mescaline block the uptake of neurotransmitters from the synaptic cleft. Depressants such as alcohol, barbiturates, and diazepam enhance inhibitory receptors, and strychnine blocks them. Atropine, curare, and hemlock block excitatory receptors, causing paralysis. (1,000,000 ×)

rum), a plant long used by mystics because of its man-shaped root. Most of us have been poisoned by atropine while having our eyes checked at the ophthalmologist. A small drop of atropine in each eye temporarily paralyzes the muscles controlling the iris, dilating the pupils.

Poison hemlock (*Conium maculatum*) contains the poisonous substance coniine, which paralyzes muscles in the same way as atropine and curare. Hemlock is a common plant in Europe and Asia, now naturalized in North America, which strongly resembles nonpoisonous Queen Anne's lace, carrots, and celery. Socrates is the most celebrated historical victim of hemlock poisoning. Descriptions of his death show the typical course of hemlock poisoning. He reportedly walked for about a quarter of an hour after drinking the poisoned beverage, then sat when his limbs became too heavy. Death followed in the course of an hour, as the poison reached his heart.

Strychnine is a particularly noxious poison which attacks a different site than the previous three. It is obtained from a species of bush rope (*Strychnos nux-vomica*), native to India. Strychnine blocks the inhibitory receptors in nerve synapses that normally damp out the firing of nerves so that nerves do not repeatedly stimulate themselves or the surrounding nerves. Strychnine blocks this control, with a disastrous result.

Strychnine is incredibly potent: the lethal dose is a mere 15–30 milligrams. Strychnine poisoning shows a characteristic course of symptoms. Minutes after eating the tiny lethal dose, the victim begins to feel anxious and starts to twitch. In 10–20 minutes, convulsions begin. The slightest movement, sound, or stimulus causes every muscle in the victim's body to simultaneously contract, stiffening the entire body into a rigid arc. Since the spread of signals to the muscles is completely unregulated, a signal to any nerve quickly spreads in a wave to all nerves, which in turn signal all of the muscles to contract. When being treated for strychnine poisoning, a patient is kept in a dark room in very loose clothing, to reduce the sensory stimuli that would precipitate a convulsion.

Diazepam (Valium), barbiturates, and alcoholic beverages, collectively known as *depressants*, also act at the inhibitory receptor. They are thought to affect the receptor in the opposite way, however, by binding at a different site on the protein than strychnine and eliciting a more subtle response. Depressants appear to facilitate the binding of neurotransmitters to inhibitory receptors, augmenting their effect and slowing the firing of nerves. The effects of different depressants are additive—alcohol will greatly enhance the effects of barbiturates—because they all act on the same protein.

Depressants affect nerves everywhere in the body. With increasing dose, the effects progressively mount, starting with relief from anxiety and progressing to disinhibition, to sedation, to hypnosis (sleep), to general anesthesia, to coma, and finally to death. The initial few steps of this progression are the effects we desire when having a few drinks, the last few the consequences of addiction and overdose.

Alcohol is perhaps the most widely used of drugs. It has been consumed throughout recorded history and has been claimed to cure nearly every ailment known to man. Alcohol is highly caloric. Ninety-five percent of the alcohol we drink is metabolized through the citric acid cycle and oxidative phosphorylation, forming ATP. The other 5% is excreted in urine and through the breath. But alcohol, as opposed to a balanced diet, is completely non-nutritive, yielding only calories, not resources for growth and maintenance.

Amphetamines have an opposite effect than depressants. They act as *stimulants* that enhance the firing of neurons. Amphetamines are structurally similar to the neurotransmitters norepinephrine and dopamine. Amphetamines seep into the nerve cell and, because they look so much like these neurotransmitters, displace them out of the storage vesicles. The neurotransmitters then freely leak out of the cell and activate the receptors on the other side of the synapse. Amphetamines thus cause nerve cells to fire without a normal stimulus.

Psychedelics, as you might expect, also act at nerve synapses. Psilocin (the drug from mushrooms), mescaline (the drug from peyote), and LSD all are chemically very similar to the neurotransmitter serotonin. Serotonin is used in only a small fraction of synapses, some of which are directly involved in the processing of sensory inputs. Psychedelics appear to slow the removal of serotonin from the synapse, so these nerve cells remain active longer. This explains a common psychedelic experience: the crossover of senses. Colors are interpreted as tactile stimuli and vice versa. LSD is by far the strongest of the psychedelics, requiring only micrograms for action. It was discovered in 1938 by Albert Hofmann, who experienced its effects after accidentally eating a small amount. (We owe many chemical advances to careless chemists, such as the discovery of sweeteners and flavorings, and, unfortunately for their discoverers, some poisons.) He went on to explore its effects in a number of controlled experiments, using himself as the test subject.

PAINKILLERS

Morphine from the opium poppy (*Papaver somniferum*) is the most potent painkiller known (Fig. 10.4). When morphine and its derivatives such as heroin (collectively called opiates) bind to receptors in the brain, we perceive pain differently. The pain is still felt, but it is not interpreted as uncomfortable. Recent research has uncovered the enkephalins, natural molecules in our bodies acting at these receptors that transiently modify pain signals to increase our threshold for pain when needed.

Acetyl salicylic acid (aspirin) and acetaminophen (Tylenol) act at the other end of the pain process, where pain signals originate. The cells in our skin and other tissues release small molecules called prostaglandins when damaged. Elevated levels of prostaglandins signal pain nerves to fire. These drugs block the two enzymes that make prostaglandins from a common lipid. Acetamino-

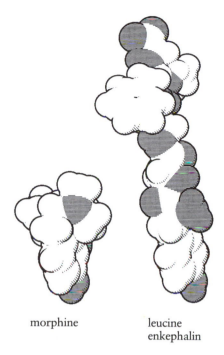

morphine   leucine
                  enkephalin

Figure 10.4   Opiate Painkillers

Morphine is chemically similar to a portion of enkephalin, a very small protein that
acts as one of our body's natural painkillers. (30,000,000 ×)

phen blocks the first enzyme, prostaglandin synthetase, and aspirin blocks the
second enzyme, cyclooxygenase.

ANTIBIOTICS

It may seem strange to us now, but the concept that microorganisms cause
disease—that a physician's unclean hands may directly result in an infected
patient—has only been understood in the past century. With the discovery of
microscopic germs, the importance of killing the organisms on the hands of
doctors and the surfaces of medical instruments became clear. The search for
specific bacterial poisons began.

The first approach, still used today, was to use an antiseptic. Antiseptics are
mild poisons that kill microorganisms by completely engulfing them. Since we
only contact the antiseptic on our skin, only a few surface cells die, not our

entire body. Solutions that contain chlorine, bromine, or iodine are effective disinfectants. These reactive atoms attack sulfur atoms in bacterial enzymes and inactivate them. Mercury compounds, such as those in merthiolate, act similarly. Alcohol is also widely used. When engulfed in concentrated alcohol, bacterial proteins unfold and are inactivated.

The next great advance in medical biochemistry was the discovery of sulfa-nilamides. These were the first substances that could be taken internally to preferentially kill microorganisms, at minimal risk to the patient. Sulfanil-amides attack a key bacterial enzyme in the synthesis of nucleotides. Even more effective antibiotic drugs were later discovered. These antibiotics were substances normally secreted by molds to control the growth of bacteria around them. The most familiar is penicillin, which blocks an enzyme that links up the sugars in the peptidoglycan layer of bacteria (Fig. 10.5). This layer provides the major support for the cell wall: without it, the cell is easily destroyed. The key to penicillin is that it attacks a molecular machine which is not used in other organisms, including ourselves. The drug is completely spe-cific to bacteria.

In what other ways are bacterial cells different from our cells? Processes that are different would provide excellent targets for antibiotic action. One difference is apparent in the pictures of Part II: bacterial ribosomes are different from our ribosomes. Chloramphenicol is an antibiotic that attacks bacterial

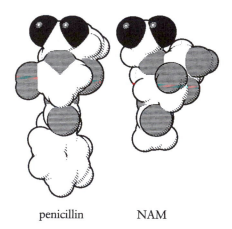

penicillin   NAM

Figure 10.5   Penicillin

Penicillin is chemically similar to the bacterial sugar N-acetyl muramic acid (NAM) and blocks the bacterial enzyme that links NAM to the growing bacterial cell wall. (30,000,000 ×)

ribosomes but does not affect the larger ribosomes in our cytoplasm. It is rather toxic, however, presumably because it also attacks the bacteria-like ribosomes in our mitochondria. Streptomycin is more specific: it attacks only bacterial ribosomes.

Unfortunately, the process of evolution is incredibly fast in bacteria, because they reproduce so quickly. Resistant strains of many bacteria have evolved in the decades since the advent of antibiotics. Bacteria gain penicillin resistance by making an enzyme (a penicillinase) that destroys the drug. Through mutation, many bacteria have developed such an enzyme, by modifying and existing digestive enzyme. In response to this, the drug manufacturers must find a slightly modified drug, different by a few atoms, which will still block the formation of the cell wall but which cannot be broken by the penicillinase. This tug-of-war is constantly underway in modern medicine.

DRUG DISCOVERY AND DEVELOPMENT

A disappointingly hit-or-miss method of finding new therapeutic drugs has been used in the past and is for the most part still used today. The story of ether is a good example. In the early days of medicine, it was found, by accident, that ether would numb a patient's senses, allowing painful surgery to be performed with much reduced discomfort. This prompted a search for better anesthetics. This search was performed in a frightfully dangerous manner: test subjects were given bowls of solvent to breathe and asked to report their effect. Because very little was known about the molecular processes underlying the numbing, substances were tried at random.

Today, more is known about the molecular processes of drug action, so the discovery and improvement of drugs is more directed. The typical approach now is to start with an existing drug and to make hundreds of small molecular changes, adding a few atoms here or there. Each of these similar drugs is then laboriously tested. They are tested in cell cultures, then the best are tested in animals, and finally the best and safest are tested in humans. The original drugs, however, are still often found entirely by accident, stumbled upon during the study of some obscure mold or plant.

The expanding field of computer-aided drug design shows great promise for accelerating and streamlining this process in the future. The idea is to test thousands of possible drugs in the computer, modeling the way that the drug will interact with the target molecular machines. The computer testing should be able to eliminate a large fraction of the candidate drugs by showing that they do not interact favorably with the target. Chemists and medical researchers could then synthesize and test only the best candidates.

Biochemists currently have the two pieces of information necessary for this process. First, many of the target molecular machines are known in atomic detail. Researchers have determined the exact atomic structure of dozens of proteins, and hundreds more are currently under study, providing a wealth of potential targets. Second, the forces involved in the binding of drugs to these molecular machines are well understood and may be accurately modeled in the computer. Together, this knowledge makes computer-aided drug design a reality. Of course, there are still many small kinks to be ironed out, but a sturdy foundation is in place.

PROSPECTS

Today, biochemistry is a field filled with excitement and promise. Biochemical research is taking a quantum step forward, from the realm of observation into the realm of application. Biochemistry has traditionally been an observational science. By looking at whole cells, at pieces of cells, and at the individual molecules in cells, biochemists have accumulated an enormous body of information in the past hundred years. The illustrations in this book continue in this tradition of observation: they combine a diversity of biochemical facts, gathered by innumerable scientists, into a few coherent pictures.

But just as physicists in the past century have used their observations of lightning and planetary motions to build transistors and rocket ships, biochemists are now putting their body of knowledge to work on the leading problems of medicine and industry. Biochemists are working to design drugs to attack the AIDS virus and to find methods of recognizing and destroying cancer cells. Researchers in the private sector are engineering strains of bacteria and yeast to make huge quantities of useful proteins, such as hormones, synthetic blood, and cancer-fighting agents. Molecular biologists routinely edit, splice, and re-arrange the DNA in cells, designing pest-resistant plants or pollutant-eating bacteria. Our growing understanding of the machinery of life is giving us the power to control and improve the biological aspects of our lives.

Further Reading

TEXTBOOKS ON BIOCHEMISTRY

Alberts B, Bray D, Lewis J, Raff M, Roberts K, Watson JD. *Molecular Biology of the Cell*. New York: Garland, 1989.

Stryer L. *Biochemistry*. New York: Freeman, 1988.

SPECIALIZED TOPICS

Atkins PW. *Molecules*. New York: Scientific American, 1987.

Bosco D. *The People's Guide to Vitamins and Minerals, from A to Zinc*. Chicago: Contemporary Books, 1980.

de Duve C. *A Guided Tour of the Living Cell*. New York: Scientific American, 1984.

Dickerson RE, Geis I. *The Structure and Action of Proteins*. New York: Harper and Row, 1969.

Kessel RG, Shih CY. *Scanning Electron Microscopy in Biology*. Berlin: Springer-Verlag, 1976.

Lentz TL. *Cell Fine Structure*. Philadelphia: WB Saunders, 1971.

Neidhardt FC, Ingraham JL, Schaechter M. *Physiology of the Bacterial Cell*. Sunderland, MA: Sinauer Associates, 1990.

Roland JC, Szollosi A, Szollosi D. *Atlas of Cell Biology*. Boston: Little, Brown and Company, 1977.

Snyder SH. *Drugs and the Brain*. New York: Scientific American, 1986.

Stevens SD, Klarner A. *Deadly Doses: A Writer's Guide to Poisons*. Cincinnati: Writer's Digest, 1990.

Index